War Stories from the Drug Survey

The primary data driver behind US drug policy is the National Survey on Drug Use and Health. This insider history traces the evolution of the survey and how the survey has interacted with the political and social climate of the country, from its origins during the Vietnam War to its role in the war on drugs. The book includes firsthand accounts that explain how the data were used and misused by political leaders, why changes were made in the survey design, and what challenges researchers faced in communicating statistical principles to policymakers and leaders. It also makes recommendations for managing survey data collection and reporting in the context of political pressures and technological advances.

Survey research students and practitioners will learn practical lessons about questionnaire design, mode effects, sampling, nonresponse, weighting, editing, imputation, statistical significance, and confidentiality. The book also includes common-language explanations of key terms and processes to help data users understand the point of view of survey statisticians.

JOSEPH GFROERER was responsible for analysis and supervision of the National Survey on Drug Use and Health for more than three decades as a statistician at the National Institute on Drug Abuse (NIDA) and the Substance Abuse and Mental Health Services Administration (SAMHSA). A widely recognized expert in methods for substance use surveys, he authored dozens of peer-reviewed journal articles and book chapters and hundreds of government reports on survey methodology and substance use epidemiology. A member of the American Statistical Association for more than thirty-five years, he has received numerous awards from NIDA, SAMHSA, the White House, and the American Public Health Association for his work on the survey.

War Stories from the Drug Survey

How Culture, Politics, and Statistics Shaped the National Survey on Drug Use and Health

---

Joseph Gfroerer
US Department of Health and Human Services (retired)

CAMBRIDGE
UNIVERSITY PRESS

CAMBRIDGE
UNIVERSITY PRESS

University Printing House, Cambridge CB2 8BS, United Kingdom

One Liberty Plaza, 20th Floor, New York, NY 10006, USA

477 Williamstown Road, Port Melbourne, VIC 3207, Australia

314–321, 3rd Floor, Plot 3, Splendor Forum, Jasola District Centre, New Delhi – 110025, India

79 Anson Road, #06–04/06, Singapore 079906

Cambridge University Press is part of the University of Cambridge.

It furthers the University's mission by disseminating knowledge in the pursuit of education, learning, and research at the highest international levels of excellence.

www.cambridge.org
Information on this title: www.cambridge.org/9781107122703
DOI: 10.1017/9781316388563

First published 2019

Printed and bound in Great Britain by Clays Ltd, Elcograf S.p.A.

A catalogue record for this publication is available from the British Library.

Library of Congress Cataloging-in-Publication Data
Names: Gfroerer, Joseph C., 1954- author.
Title: War stories from the drug survey : how culture, politics, and statistics shaped the national survey on drug use and health / Joseph Gfroerer.
Description: 1 Edition. | New York : Cambridge University Press, 2018.
Identifiers: LCCN 2018026583| ISBN 9781107122703 (hardback) | ISBN 9781107553453 (paperback)
Subjects: LCSH: National Survey on Drug Use and Health (U.S.) | Health surveys–United States. | Drug control–United States. | BISAC: SOCIAL SCIENCE / Statistics.
Classification: LCC HV5825 .G48 2018 | DDC 362.290973–dc23
LC record available at https://lccn.loc.gov/2018026583

ISBN 978-1-107-12270-3 Hardback

For Sue and Rachel

Contents

Figures

Tables

Preface

In the fall of 2013, I decided to end my federal career after thirty-seven years as a statistician in the US Department of Health and Human Services. During my final months before retiring in January 2014, it occurred to me that the project I had worked on for the past thirty years, the National Survey on Drug Use and Health (NSDUH), had an interesting history, including amusing stories and valuable lessons for statisticians and government leaders. But the stories were not only about statistics and survey research; they were also about management, how government operates, politics, personalities, and the nation's drug abuse policies. I felt that this history would be of interest to a broad audience, not just survey researchers. I also knew that these stories from years ago were still relevant because they were often used as examples and justification to guide current decision-making, or simply to explain why the survey was the way it was. I realized that the only way this history would be appreciated and preserved was for me to write the story. My direct involvement in the survey since the early 1980s, including serving as the lead federal official responsible for managing the project from 1988 through 2013, gives me a unique perspective on the survey's history. I had saved much of the survey's documentation in my paper and electronic files, and also in my head. With the aid of the collection of published and unpublished reports, internal memos, notes from meetings, and interviews with other people involved in the survey, I was able to construct a complete chronicle of the survey. Most of it is based on my firsthand knowledge of the events described. Keeping in mind the wide range of people who may be interested in learning about how surveys are conducted, drug policy, and government, I have kept complex statistical discussions to a minimum. There are no formulas in the book, just simple explanations of some key statistical concepts.

My initial work on the survey was at the National Institute on Drug Abuse, conducting analysis with the data files from the 1974–79 surveys. I participated in planning for the design of the 1985 survey. I became alternate project officer in 1983, and project officer in 1988. With full responsibility for managing the survey contract, and little staff support, it was necessary to become familiar with every aspect of the project. As the survey grew in size and importance, and moved to the Substance Abuse and Mental Health Services Administration, I was able to gradually recruit and hire staff with a wide range of survey-related expertise to build a strong, diverse team to manage the project. The survey team has faced many difficult management, design, and analysis problems. The solutions we implemented often worked but sometimes failed. These experiences serve as lessons that can guide statisticians and survey managers in their work, and suggest factors that are associated with survey success. I am pleased to share these experiences with other statisticians and managers of surveys, to help them make sound decisions when they face similar challenges.

Joseph Gfroerer
Frederick, Maryland
August 2018

Acknowledgments

Special thanks go to Joe Gustin, who helped manage the survey contract from 1989 until 2006. He brought his passion for history to work with him, which triggered his idea to draft a report on the early history of the survey, focusing on contracting and contractors. The information he gathered, including documentation and interviews with early NSDUH leaders Herb Abelson, Louise Richards, Joan Rittenhouse, and others, was invaluable in writing the first few chapters of this book.

I thank Tim Johnson for his encouragement and advice to me as I developed the concept of the book and the proposal I submitted to the publisher. He also reviewed my initial drafts of early chapters.

I am hugely indebted to Jonaki Bose for the time she devoted to reviewing drafts of this book. She reviewed every chapter (sometimes second and third drafts), providing valuable technical comments and suggestions that without a doubt improved the book.

Others who helped by reviewing drafts of portions of the book, locating and sending me reference documents, and talking to me about their experiences with the survey include Edgar Adams, Peggy Barker, Ann Blanken, John Carnevale, Judy Droitcour, John Gfroerer, Sarra Hedden, Art Hughes, Joel Kennet, Anna Marsh, Grace Medley, Dicy Painter, Coleen Sanderson, Len Saxe, Peter Tice, Tom Virag, Mark Weber, and Terry Zobeck.

I would also like to thank all of the great staff that worked under me on the survey from 1988 to 2014, when I retired. All were dedicated and productive, and the survey's successes are due to their work. The project benefited greatly from staff who stayed with the survey for a long time, building their in-depth knowledge of the survey and institutional memory. I list them all here, grouped by the length of time they worked on the NSDUH team, as of 2017. Peggy Barker, Joe Gustin, Art Hughes, Joel Kennet, Dicy Painter, and Doug Wright all devoted fifteen or more years to the NSDUH. Jonaki Bose, Joan Epstein, and Pradip Muhuri contributed more than ten years. Those with fewer than ten years on the team were Marc Brodsky, Jim Colliver, Lisa Colpe, Janet Greenblatt,

Beth Han, Lana Harrison, Sarra Hedden, Mike Jones, Andrea Kopstein, Sharon Larson, Rachel Lipari, Grace Medley, Jeanne Moorman, Ken Petronis, Kathy Piscopo, Maria Rivero, Lucilla Tan, and Pete Tice. In addition, the success of the survey would not have been possible without the great work of RTI's contract managers, task leaders, statisticians, survey methodologists, programmers, field supervisors, field interviewers, and others. There are too many to name, but special thanks go to Tom Virag, who served as RTI's project manager on the main survey contract from 1988 until he retired in 2014. I also appreciated the excellent work of the NORC staff on the NSDUH analysis contract.

My greatest thanks go to my wife Sue, who was supportive of my frequent weeknight and weekend work and the on-call nature of my responsibility for overseeing NSDUH during my HHS career and in the early years of my "retirement." She reviewed drafts of every chapter, and the book was made more readable because of her editing skills and her non-statistician perspective.

Acronyms

ACASI	audio computer-assisted self-interviewing
ADAMHA	Alcohol, Drug Abuse, and Mental Health Administration
ASPE	Office of the Assistant Secretary for Planning and Evaluation, HHS
CAI	computer-assisted interviewing
CAPI	computer-assisted personal interviewing
CBHSQ	Center for Behavioral Health Statistics and Quality
CDC	Centers for Disease Control and Prevention
CIPSEA	Confidential Information Protection and Statistical Efficiency Act
CMHS	Center for Mental Health Services, SAMHSA
CODAP	Client Oriented Data Acquisition Process
CSAP	Center for Substance Abuse Prevention, SAMHSA
CSAT	Center for Substance Abuse Treatment, SAMHSA
DAWN	Drug Abuse Warning Network
DC-MADS	DC Metropolitan Area Drug Study
DDID	Division of Data and Information Development, NIDA
DEA	Drug Enforcement Administration, Department of Justice
DEPR	Division of Epidemiology and Prevention Research, NIDA
DESA	Division of Epidemiology and Statistical Analysis, NIDA
DMPA	Division of Medical and Professional Affairs, NIDA
DPS	Division of Population Surveys, OAS
DUF	Drug Use Forecasting
FI	field interviewer
FTE	full time equivalent
GAO	General Accounting Office
GWU	George Washington University
HEW	Department of Health, Education and Welfare

HHS	Department of Health and Human Services
ISR	Institute for Survey Research, Temple University
LA	listing area
MHSS	Mental Health Surveillance Study
MTF	Monitoring the Future study
NCHS	National Center for Health Statistics
NDATUS	National Drug and Alcoholism Treatment Unit Survey
NFIA	National Families in Action
NHIS	National Health Interview Survey
NHSDA	National Household Survey on Drug Abuse
NIAAA	National Institute on Alcohol Abuse and Alcoholism
NIDA	National Institute on Drug Abuse
NIH	National Institutes of Health
NIJ	National Institute of Justice
NIMH	National Institute of Mental Health
NOMS	National Outcome Measures
NORML	National Organization for the Reform of Marijuana Laws
NSDA	National Survey on Drug Abuse
NSDUH	National Survey on Drug Use and Health
OAS	Office of Applied Studies
ODAP	Office of Drug Abuse Policy
OMB	Office of Management and Budget
ONDCP	Office of National Drug Control Policy
PAPI	paper-and-pencil interviewing
PART	Program Assessment Rating Tool
PDFA	Partnership for a Drug Free America
PRIDE	Parents' Resource Institute on Drug Education
PSU	primary sampling unit
RAC	Response Analysis Corporation
R-DAS	Restricted Use Data Analysis System
RFP	Request for Proposal
RTI	Research Triangle Institute (RTI International)
SAE	small area estimation
SAMHSA	Substance Abuse and Mental Health Services Administration
SAODAP	Special Action Office for Drug Abuse Prevention
SAPT	Substance Abuse Prevention and Treatment
SBIRT	Screening, Brief Intervention, and Referral to Treatment
SED	serious emotional disturbance

SMI	serious mental illness
SPG	Special Projects Group
SSDP	State Systems Development Program
SSR	state sampling region
TEDS	Treatment Episode Data Set
YARM	yet another redesign meeting

Introduction

This book tells the story of a survey. Not just any survey, but a very big survey, called the National Survey on Drug Use and Health (NSDUH).[1] It is one of the largest ongoing surveys conducted by the US federal government, and is the nation's principal source of data on illicit drug use among the US population. The survey has a long and interesting history, involving scientific controversy, arguments over the design and funding of the survey, political grandstanding, important research findings, and occasional embarrassing mistakes. The survey started nearly fifty years ago as a small research study collecting data from just over 3,000 randomly selected respondents, at a cost of $211,500. Since then, the survey has expanded in size, scope, and utility, reaching an annual cost of nearly $50 million and interviewing almost 70,000 Americans each year. You may have seen news accounts reporting the results of the survey over the past four decades. Here are some of the headlines:

1980
"Reports show dramatic increase in use of marijuana and cocaine"[2]

1990
"Bush Hails Drug Use Decline in a Survey Some See as Flawed"[3]
"Senator: Survey 'wildly off the mark'"[4]

2000
"Colorado leads U.S. in marijuana use"[5]
"Massachusetts worst in drug use, survey finds"[6]
"Delaware leads U.S. in teen drug use"[7]

2009 and 2010
"New National Survey Reveals Significant Decline in the Misuse of Prescription Drugs"[8]
"National survey reveals increases in substance use from 2008 to 2009; Marijuana use rises; prescription drug abuse and ecstasy use also up"[9]

2014 and 2015
"More Americans are using marijuana"[10]
"Teen drug and alcohol use continues to fall, new federal data show"[11]
"Heroin use surges, addicting more women and middle-class"[12]
"Teen pot use holds steady in first year of legal weed, new federal data show"[13]

The headlines illustrate how government leaders and the media understood and communicated the findings from the survey. Of course, headlines don't give the whole story, but these brief snippets are telling. They exhibit disagreements on the interpretation of results, self-serving statements, and contradictory findings, as well as actual shifting patterns of drug use. The headlines trigger a host of questions. How does the government come up with these numbers? How is the survey conducted, and do the survey managers really believe private citizens willingly tell the government about their illegal drug use? Who decides what kinds of data the survey collects, and from whom? How can survey participants be sure that the information on their illegal activities and other personal information is not shared with law enforcement, employers, or others? Do government officials report the data objectively, or do they "spin" it to promote their own political agendas or preferred policies? Does the government actually use these data to develop policies and programs? These fundamental questions have been raised by government leaders, researchers, reporters, and the public for decades. One goal of this book is to provide answers to these questions, in the context of specific events that occurred throughout the history of the survey.

The book tracks the changes in the design of the survey and the way the results were reported, explaining how these changes were influenced by cultural, political, personal, and statistical concerns. External events that influenced the survey include the Vietnam War, overdose deaths of famous athletes, and states passing legislation legalizing medical and recreational marijuana use. The goals and content of the survey shifted when different divisions or agencies gained control over the project. Frequently, but perhaps less prominently, the survey was affected by technical, scientific concerns and associated attempts to improve the survey methods.

A principal focus of the book is the important role of science in the success of surveys. Science in this context specifically refers to the established principles of the field of survey research and statistics. Following these principles leads to statistical integrity, which refers to the respect and trust people have for the survey staff and the data they produce. The evolution of NSDUH from a small periodic research study to a multimillion-dollar ongoing survey that became the nation's leading barometer of trends and patterns of substance abuse in the population is largely attributable to the recognition of and adherence to these principles. But the path has not always been smooth. Throughout NSDUH's history, there have been many examples of conflicts, decisions, successes, and failures associated with efforts to produce high quality, useful data while maintaining statistical integrity.

Statistical integrity involves exhibiting a strong commitment to statistical rigor, transparency, and unbiased reporting of results. In Chapter 10 the book discusses our efforts to objectively report the survey's results by retaining maximum control over the timing, content, and interpretation of new data releases. A full report on the results and methods, including limitations and caveats associated with the data, was released each year at the regularly scheduled kickoff event for Recovery Month. The report was prepared by the NSDUH staff, with no substantive review and revision by political leaders.

An aspect of unbiased reporting is resisting and speaking out against inappropriate uses of data and poor survey methodologies. Of course, politicians cannot always be trusted to objectively report the survey results. Chapters 3 through 6 describe politically motivated interpretations by drug czars, despite the straightforward, objective publications the survey team produced. The survey team has shown resistance to these types of distortions, starting with the first drug survey, in 1971. Chapter 1 explains President Nixon's urging to have the report containing the 1971 survey results emphasize problems caused by marijuana use. Nevertheless, when the report was released it highlighted findings that marijuana did not pose a major public health threat, and the public perceptions about the dangers of marijuana were unfounded. In another case, described in Chapter 8, President Clinton used NSDUH data to claim that a decline in drug use in Miami was evidence of the success of prevention efforts, despite a NSDUH report in progress (and published after his Miami announcement) that concluded the decrease was likely an artifact of the effects of Hurricane Andrew on the sample.

Effective communication is critical for the success of a survey, as demonstrated by numerous examples in the book. Statisticians must be able to explain to data users some of the technical aspects of the survey, such as how the sample was selected, how data were collected, procedures for making estimates, and the caveats associated with results. This requires special skill in translating complex statistical concepts into descriptions that are understandable to non-statisticians. Communication failures can result if survey staff are not sensitive to the areas of expertise of the people they are communicating with. Chapter 5 describes a situation in which the National Institute on Drug Abuse (NIDA) was criticized for taking three months to inform the White House Office of

National Drug Control Policy (ONDCP) about an error discovered in previously published, politically sensitive estimates of heavy cocaine users. The delay was due to our difficulties in explaining statistical aspects of the error to NIDA's director. Chapter 4 describes the planning for a methodological study that went awry because of a simple misunderstanding of the term "nonresponse bias" by staff at ONDCP. Delays and wasted effort could have been avoided by having an initial meeting between statisticians and ONDCP to discuss the goals of the study.

Effective survey management must include appropriate communication and coordination within the project staff, across the different groups responsible for aspects of the survey, such as sampling, data collection, processing, and reporting results. Important quality control processes and discoveries resulted from the establishment of links between experts within the NSDUH team. Report writers worked with data processing statisticians to create a system to flag estimates unduly affected by editing and imputation. Analysts working with field managers were able to detect that the experience level of interviewers affected respondent reporting of drug use. Major redesigns of the survey described in Chapters 9 and 12 were developed in coordination with staff responsible for each project component.

An ongoing program of methodological studies to evaluate data and make improvements to survey processes should be an integral part of any large survey program, as it has been for NSDUH. Results of these studies have identified data problems, verified survey findings, and guided the development and implementation of survey design improvements. The Clinton administration's decision to expand the NSDUH in 1999 to provide data for every state, discussed in Chapter 9, was influenced by our 1996 methodological study that showed the feasibility of a small area estimation model that could produce state estimates without the need for a large sample in every state.

Throughout the survey's history, outside consultants have frequently been asked to participate in planning and decision-making on the project. The contributions of these highly regarded experts in survey design, substance use research and policy, and other NSDUH-relevant areas are mentioned in most chapters of the book. Soliciting advice from outside experts and data users is critical to the success of a large scale survey. Besides simply giving us their helpful ideas, their endorsements of our proposed plans facilitated approvals of those plans by agency heads and other decision-makers.

But there are limits to how much a survey program can rely on external consultants for directing a project. Outside experts generally will not have in-depth knowledge of the survey, and may have particular points of view or self-interests that don't line up with the agency goals for the survey. Ultimately, it is the survey staff that is responsible for the day-to-day operation of the survey. It's essential that this staff have the background and expertise in various areas relevant for the project, such as sample design, questionnaire design, data collection methods, and statistical analysis. It's also important for staff to have knowledge of the subject matter of the survey and the policy and research questions that data from the survey should address. While most of the manpower on a project might be contractor staff, it is still critical to have sufficient in-house staff who are experts in these fields to manage a large project like NSDUH. This has always been a challenge. Chapter 6 describes the negotiations surrounding the transfer of the survey from the National Institute on Drug Abuse (NIDA) to the Substance Abuse and Mental Health Services Administration (SAMHSA) in 1992. NIDA initially proposed that SAMHSA would need only one person to manage the project, but increased it to three during negotiations. Although SAMHSA added staff for NSDUH over time, reductions and reorganizations beginning in 2005, discussed in Chapter 11, had detrimental effects on the survey and staff morale.

A survey cannot be considered a success without a strong record of producing relevant, informative results. Besides summarizing the annual reports of the NSDUH results, this book describes studies that focused on specific substance use issues of interest. These include studies estimating heroin use and addiction, including links to misuse of prescription pain relievers; studies estimating how many people need treatment for substance use problems; studies to predict future substance abuse treatment need; studies of recent trends in drug abuse among aging baby boomers; and an analysis of drug use among women prior to pregnancy, during pregnancy, and after childbirth. The book also describes efforts to make the NSDUH microdata files available to researchers outside the survey team, resulting in hundreds of studies published in professional journals.

The story of the survey is told chronologically. Each chapter covers a broad era of the survey's and the nation's history, as they are deeply intertwined. The specific events, debates, and decisions that occurred during each phase of the survey's history are described in the first twelve chapters. Brief discussion narratives that focus on recurring themes of the book are inserted, following Chapters 2, 5, 8, 10, and 12. A final chapter includes conclusions and discusses future considerations. The Appendix contains tables that give a concise overview of the history of the survey, including contractors, sample design, and response rates.

Other Histories of Government Surveys

This book adds to the considerable literature documenting the development of the US federal statistical system. A broad overview of the early history of the entire system, was published in 1978.[14] It covers many of the same themes as this book, such as probability sampling, the impact of technical developments such as computerization, political and legislative events impacting surveys, organization and coordination of data programs across agencies, statistical integrity, confidentiality, and the use of advisory committees. Other relevant works include histories of the Bureau of Labor Statistics,[15] the Decennial Census,[16] and the Current Population Survey.[17] Some studies have focused on the development of important official measures such as poverty,[18] unemployment,[19] and race,[20] and how these measures have evolved over time. This book briefly touches on difficulties and decisions regarding different measures associated with substance abuse, such as the overall level of drug use, heroin use, treatment need, recovery, and drug consumption.

Who Should Read this Book

Although a basic knowledge of statistics and survey research will be helpful to readers of the book, it is not a requirement. My goal was to make this story accessible and interesting to a wide range of readers, including survey statisticians, other researchers, policymakers, leaders of government and private organizations that conduct surveys, journalists, and the general public with an interest in drug abuse policy and history. Where possible I have included simple explanations of key terms and processes associated with statistical methods and survey research. This approach is consistent with a recurring theme of the story, one reason that I wrote the book: There is a need for better understanding and communications between the statisticians who conduct surveys and the program managers and policymakers who ask for surveys and use the data. This book should also be useful to students of statistics and survey design, providing descriptions of real-life experiences in the development of survey designs, management of surveys, and analytic approaches. The focus is on a large, ongoing government survey, but most of the examples and lessons discussed are relevant for any survey, regardless of size or sponsor. The book will help students understand the factors that must be considered in survey research, beyond the material covered in standard textbooks.[21]

Notes

1 The survey was given this name in 2002. Prior to 2002, the survey names had been National Household Survey on Drug Abuse (NHSDA) from 1985 to 2001, National Survey on Drug Abuse (NSDA) from 1977 to 1982, and informally the National Survey or the Household Survey. All of these names are used throughout the book.
2 HHS News, June 20, 1980
3 Treaster, Joseph, *New York Times*, December 20, 1990, B14.
4 Kelly, Jack, *USA Today*, December 20, 1990.
5 Guy, Andrew, *Denver Post*, September 1, 2000, A1.
6 Donnelly, John, *Boston Globe*, September 1, 2000, A1.
7 Church, Steven, *News Journal*, Wilmington, DE, September 2, 2000, A1.
8 SAMHSA press release, September 10, 2009.
9 SAMHSA press release, September 16, 2010.
10 *USA Today*, September 5, 2014.
11 *Washington Post*, Wonkblog, Sept. 16, 2014.
12 Szabo, Liz, *USA Today*, July 7, 2015.
13 *Washington Post*, Wonkblog, September 10, 2015.
14 Duncan and Shelton, *Revolution in United States*.
15 Goldberg and Moye, *The First Hundred Years*.
16 Anderson, *The American Census*.
17 Bregger, "The Current Population Survey."
18 Ruggles, *Drawing the Line*.
19 Card, *Origins of the Unemployment Rate*.
20 Prewitt, *What is Your Race*.
21 Fowler, *Survey Research Methods;* Groves et al., *Survey Methodology*.

1 President Nixon Launches the War on Drugs

Historians have pointed out the significance of the events of 1968. The Vietnam War was raging. The Tet Offensive launched by the North Vietnamese forces in January and February gave warning to Americans that victory would not be certain or soon.[1] Public opinion turned against the war. According to the Gallup Poll in February of that year, 35 percent of Americans approved of President Lyndon Johnson's handling of the war, while 50 percent disapproved, with 15 percent having no opinion.[2] In March, anti-war candidate Eugene McCarthy surprisingly won 42 percent of the vote in the New Hampshire Democratic primary election for president, nearly defeating the incumbent president. A few days later, Johnson announced that he would not run for reelection in November. The assassination of civil rights leader Martin Luther King on April 4 triggered riots across American cities. Then on June 4, Democratic presidential candidate Robert Kennedy was assassinated. In August there were anti-war demonstrations and clashes between protesters and police at the Democratic National Convention in Chicago. In November, Republican Richard Nixon, promising to bring the war to an end and emphasizing law and order, was elected president, narrowly defeating Democrat Hubert Humphrey, Johnson's vice president.

The summer of 1969 was also an eventful time in America. Nixon had just been inaugurated in January. Anti-war demonstrations subsided in anticipation of a peace agreement. But a peace plan proposed by the North Vietnamese in May 1969 was rejected by the Nixon administration. Anti-war demonstrations resumed. By May 1969, the number of American troops in Vietnam had swollen to over 500,000. Most were under age 25. More than 40,000 had been killed, including nearly 17,000 in 1968 alone.[3] Then on June 8, 1969, President Nixon announced that 25,000 US troops were being withdrawn from Vietnam. This was the beginning of the new "Vietnamization" policy. South Vietnamese troops would assume increasing responsibility for the fighting as the United States gradually reduced its troop levels. While not an overt admission of defeat, it was a turning point in the war effort. American casualties had increased each year up to 1968 (the peak year) and then declined in 1969 and each year afterward, until a peace settlement was finally reached on January 23, 1973.

During the summer of 1969 there was a more positive historic event that demonstrated the technological prowess of the United States. On July 20, Apollo 11 landed on the moon, and Neil Armstrong became the first person to set foot on it. Millions of Americans watched that first moonwalk on their televisions at home that evening, with a sense of awe and pride in their country. I watched it with my parents in our living room.

As all this was happening, the Nixon administration was preparing to launch a new war. On July 14, just a month after troop withdrawals in Vietnam began and a week before the moon landing, the president announced a "war on drugs."[4] It was in response to the proliferation of illicit drug use by young people across the country, especially college students, and soldiers in Vietnam. A month after the announcement of the war on drugs, the Woodstock Music and Art Festival drew an estimated 500,000 mostly young people to a field in Bethel, New York, seventy miles northwest of New York City, for a weekend of fun and music (August 17–20, 1969). My brother John was there. *Time* magazine reported that most attendees used marihuana[5] at the festival.[6] Widely covered in the media, "Woodstock" became a symbolic event for a generation of young people rebelling against societal norms. Most of them opposed the Vietnam War. Many used drugs. Rock and roll was their shared music, and long hair was their trademark. "Make love not war" was a popular slogan, and the peace sign was a common greeting. Many in the Woodstock crowd probably were not even aware that the government had launched a war on drugs. But in hindsight, the stark contrast between Woodstock and Nixon's July 14 announcement made it seem like, in the war on drugs, battle lines were being drawn.

Drug Use Prior to the 1960s

It may have seemed to many Americans during the 1960s that widespread use of psychoactive drugs for recreational purposes was a new phenomenon in the country. The fact is that the United States has a long history of drug use. Opiate use was prevalent throughout the 1800s. Opium import data tracked by the federal government showed that as early as 1840, there was enough to supply a daily dose to roughly 100,000 users, or an occasional dose to millions.[7] Most of the opium was used for medical purposes. But the rate of per capita consumption doubled between 1870 and 1895,[8] with much of it suspected to be for nonmedical use.[9] Although some of that increase has been attributed to addiction stemming from widespread use of morphine on injured soldiers during the Civil War, another potential factor was the influence of the temperance movement, which began in the 1870s and advocated abstinence from alcohol. This may have created a demand for other substances among the population who felt a need for some kind of intoxicant. Increases in opiate use coincided with decreases in alcohol use. Studies showed that most opiate users in the late 1800s were female, and the typical user was in her thirties, middle and upper class, and received her first dose from a doctor.[10] Cocaine use also became prevalent in the late 1800s. By 1900, morphine, heroin, and cocaine were all legal and available,[11] and used recreationally by many Americans. These drugs were even marketed as ingredients in popular consumer products like soft drinks and cough medicines. However, the addiction and other problems caused by these drugs were becoming apparent, and laws were passed to restrict their availability. Between 1900 and 1940, their use declined dramatically. Marihuana smoking, first introduced in the United States during the 1920s, was popular among certain segments of the population but did not become widespread until the surge in the early 1960s.[12] Although there is a lack of data to accurately measure the prevalence of illicit drug use prior to the 1960s, it is clear that the levels of marihuana use reached by the late 1960s were unprecedented in American history.

Links between the Vietnam War and the War on Drugs

Without a doubt, there were links between the Vietnam War, the explosion of illicit drug use in the 1960s, the war on drugs, and consequently the birth of a national survey on drug use. Heroin became a concern when it first became cheap and widely available in Vietnam in 1969.[13] There were reports that narcotic use was common among soldiers.[14] Some officials were concerned that the returning soldiers would bring their narcotic habit back home with them, creating a surge in the need for treatment and rehabilitation.[15] The concern was so great that the federal government set up a urine testing program in 1971 to identify heroin users among servicemen leaving Vietnam, and enroll them in detox before they could return home. Later, in a study assessing the scope of the problem, data were collected from a representative sample of veterans who had returned to the United States during September 1971. The study found that nearly half had used narcotics while in Vietnam, including about 20 percent who had become addicted at some point. Of those who had ever been addicted, about half had tried narcotics again after returning home, but only 6 percent became addicted again.[16] These data

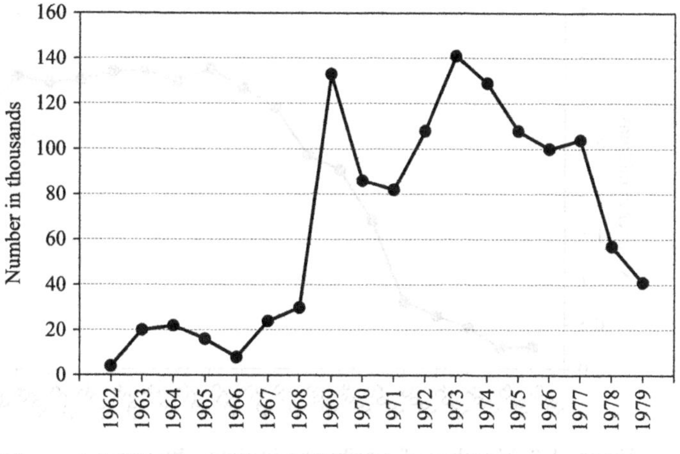

Figure 1.1 Number of heroin initiates, by year.

show that the returning servicemen did not create a surge in people needing treatment in the United States. Nevertheless, this study as well as other data suggest that the high rate of heroin use among Vietnam servicemen was a major contributing factor in the expanding use of heroin in the United States during the late 1960s and early 1970s. Figure 1.1 shows the annual US estimates of heroin initiation (first-time use), constructed from respondent reports of age at first use in later surveys.

The rate rose dramatically in 1969, and began a steady decline after 1973. Approximately one-third (35 percent) of heroin initiation among all Americans during 1969–72 occurred among soldiers in Vietnam.[17]

Concerns about marihuana use among soldiers arose early in the war. But the vast majority of US marihuana initiates during the 1960s and 1970s had never served in Vietnam. Figure 1.2 shows estimates of the annual number of marihuana initiates.

Increasing marihuana initiation was evident as early as 1964, before the war became a major public concern.[18] Anecdotally, the famous meeting in which Bob Dylan introduced marihuana to the Beatles occurred on August 28, 1964. During the war years (1964–72), marihuana initiation among Americans was increasing, but only 8 percent of the initiation occurred among servicemen in Vietnam. After the war, high rates of marihuana initiation persisted.

Thus, the evidence that drug-using Vietnam veterans caused the increases in drug use in the United States during the late 1960s and 1970s is mixed – a small impact for marihuana use, and possibly a substantial impact for heroin use.

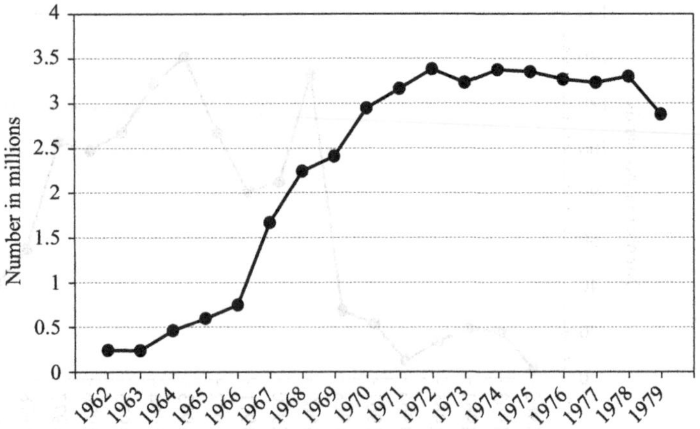

Figure 1.2 Number of marihuana initiates, by year.

The Vietnam War may have had a more subtle, indirect effect on drug use at home. During the 1960s, the growing unrest and rebellious attitude among young people grew out of a mistrust of establishment values, rules, and policies. These included racial prejudice, the war, and marihuana prohibition. As the Vietnam War expanded, its purpose was questioned, and opposition grew. Between 1965 and 1968, 1.4 million young Americans were drafted and inducted into the military. But more than 10 million 18- and 19-year-old men were at risk to be drafted. By law, every 18-year-old male was required to register with the Selective Service to be eligible to be drafted and potentially sent to fight in a war that many Americans felt was pointless. My three older brothers all registered, including one who registered in 1968, at the height of the war. I registered in 1972, the year the draft ended. Images on television and in newspapers brought the horror of the war into people's homes on a daily basis, and returning veterans shared stories with friends and family. The anti-establishment youth movement was widespread. It was symbolized by common language, dress, music, and habits, including marihuana use. Many, including my brother, participated in anti-war demonstrations. Even those youth who were not vigorous anti-war protesters or activists conformed to the cultural norms, including the dress style, long hair, rock and roll and psychedelic music, and drug use. Some well-known musicians assumed roles as leaders of the counterculture, openly using drugs, speaking out in favor of drug use, and writing drug-influenced music with drug-advocating lyrics. There were also songs about revolution and peace.

Whether or not the Vietnam War impacted youth drug use attitudes and behaviors, it is clear that President Nixon's perception of the cultural revolution influenced his plans for the war on drugs. By emphasizing the criminal aspect of drug use and maintaining the illegality of marihuana, Nixon thought the drug war could help stigmatize the youth culture of protest and rebellion, and provide a means for the anti-war crowd – his enemies – to be arrested and jailed.[19] Perhaps he was aware of the Gallup Poll of 1969 that showed 49 percent of liberal students had smoked marihuana, compared with only 10 percent among conservative students. Also, 40 percent of students who had participated in a political demonstration had used marihuana, compared with 15 percent of non-participants.[20]

The Drug War Begins

Shortly after his announcement of the war on drugs, Nixon submitted legislation to Congress for consideration. The legislation emphasized interdiction, law enforcement, and the impact of drug use on crime more than public health. Under Nixon's proposed plan, primary federal responsibility and funding would fall under the Department of Justice. Health advocates and officials at the Department of Health, Education, and Welfare (HEW) publicly objected, angering the White House. This led to the firing of Dr. Stanley Yolles, director of the National Institute of Mental Health (NIMH), although Yolles claimed to have resigned. His resignation letter criticized the administration for making the Department of Justice the "final authority in medical determinations." Joseph English, the administrator of the Health Services and Mental Health Administration (parent agency of NIMH) was also fired, and HEW Secretary Robert Finch was removed from his position as well. But the final bill that passed after more than a year of haggling did include funding in HEW for community mental health centers, research, and education programs.[21] On October 27, 1970, the president signed the Comprehensive Drug Abuse Prevention and Control Act of 1970, saying in his remarks that it was needed "because our survey of the problem of drugs indicated that it was a major cause of street crime in the United States."[22]

Title II of the Comprehensive Drug Abuse Prevention and Control Act of 1970 is the Controlled Substances Act, which placed all drugs into five categories ("schedules") according to their safety, medical uses, and abuse potential. Despite objections from the medical profession and HEW officials, marihuana was deemed by the Nixon administration as a Schedule I narcotic,[23] along with heroin and LSD. Schedule I includes drugs having no therapeutic value and maximum abuse potential. Although experts in drug abuse criticized this designation, and it was intended to be temporary pending further study, marihuana remains Schedule I to this day.[24] Coupled with the emphasis on law enforcement in the drug war, it led to millions of marihuana-related arrests over the next three decades. Of course, the scheduling of drugs impacted the content of future drug surveys. And occasionally, data from the survey were used to help the federal government make decisions about approval and scheduling of particular drugs.

The Comprehensive Drug Abuse Prevention and Control Act also established the National Commission on Marihuana and Drug Abuse. The bill initially provided $1 million to fund the commission's activities and studies, but more funds were added later. Included in the commission's charge was a study of the nature and extent of marihuana use in the United States, and its relationship with the use of other drugs. This stipulation was the genesis of the National Survey on Drug Use and Health (NSDUH).

The commission consisted of thirteen members, including two US Senators and two US Representatives. Most members were handpicked by Nixon and supportive of his drug war intentions. It was chaired by Raymond Shafer, former prosecutor and law and order Republican governor of Pennsylvania. The commission was supported by a variety of staff and consultants.[25] It was referred to as the Shafer Commission. The marihuana study began with a literature search, which identified more than 100 surveys relating to marihuana use that were conducted by schools, independent research groups, polling organizations, and governmental agencies between 1965 and 1971. Many of these surveys provided valuable information within their purview, but the diverse nature of their methodologies precluded any integration of findings. They could not satisfy the commission's desire for national estimates, so the commission decided to conduct two national surveys. The first one would cover marihuana use, and the second would address the use of other illegal drugs and their relationship to marihuana use. On August 16, 1971, the commission awarded a contract to Response Analysis Corporation (RAC) to conduct the first survey, at a cost of $211,500. The contract required the survey to be completed and the results submitted to the commission by the end of the year. RAC was a small survey research company located in Princeton, New Jersey. Dr. Herbert Abelson was its president. After gaining many years of experience as a research psychologist and vice president of Opinion Research Corporation, he formed RAC in 1969, along with Dr. Reuben Cohen. Undoubtedly, RAC was chosen to conduct this first national survey on a topic of considerable sensitivity to the general public because they had recently conducted a national survey on another sensitive topic, pornography.[26]

Design of the First National Drug Survey

In planning and designing the drug survey, Abelson worked directly with Dr. Ralph Susman, Associate Director for Sociology on the commission staff. The initial survey was officially called "A Nationwide Study of Beliefs, Information and Experience." Informally it became known simply as the National Survey or the Household Survey. The latter name became popular to distinguish it from another major national survey, the Monitoring the Future study (MTF). Sometimes called the High School Survey, MTF began in 1975 and has surveyed nationally representative samples of high school seniors every year since then, through a grant from the National Institute on Drug Abuse (NIDA) to the University of Michigan.[27]

Many of the features of the design of the 1971 and subsequent National Surveys were derived from the experience that RAC had gained in conducting the pornography survey. The survey used a multistage area probability sample, similar to designs used in many large-scale national household surveys. Like other survey organizations, RAC had developed a standing nationally representative sample of geographic areas they could use for multiple surveys. This efficiency helped them to keep costs down and also allowed them to complete a survey in a short time span. With data collection for multiple surveys carried out in the same sample areas, RAC could maintain a cadre of interviewers in those selected areas, to collect data as needed for any particular survey the firm conducted. Shown in Figure 1.3, RAC's sample for the 1971 National Survey began with a first-stage sample of 103 Primary Sampling Units (PSUs) that consisted of counties or groups of counties, selected to be representative of the coterminous United States (Alaska and Hawaii were not covered). A second-stage sample of 600 locations was selected from the 103 PSUs. These interviewing locations each had a population of approximately 2,500 persons.[28]

For the drug survey, 200 interviewing locations were sub-selected out of the full 600 in the RAC sample. An additional seventy-five interviewing locations were selected in three designated metropolitan locations (Chicago, Omaha, and Washington, DC). The next stage of sampling was the selection of one or more "segments" within each of the 275 interview locations. Segments were small areas of ten to

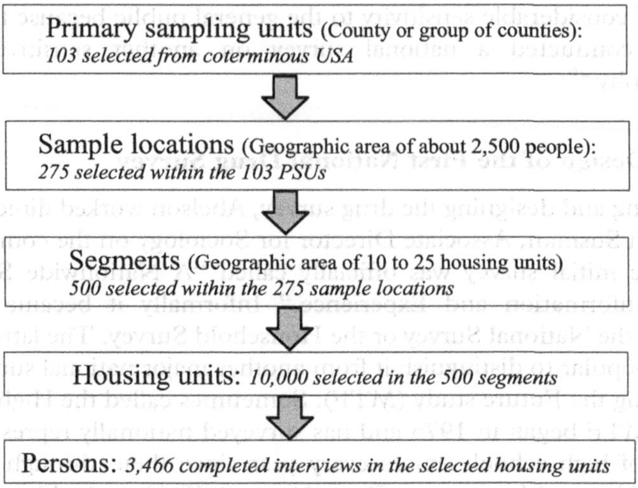

Figure 1.3 Multistage sample for 1971 survey.

twenty-five housing units. Introductory letters ("lead letters") from RAC were mailed to each of the addresses in these sampled segments. Then an interviewer visited to ask residents to complete the survey. In each housing unit, the interviewer began by making a listing of all eligible household members. Next, they used a predetermined process that randomly selected a household member or members to be asked to complete the interview. Controlled, random sampling procedures were used at each stage of sampling so that the sample, with proper weighting, would produce nationally representative estimates.

Interviewer training was conducted by eight members of the RAC study team in twenty-two separate locations throughout the United States. A contingent of 236 interviewers attended one-day sessions. Data collection began immediately after each interviewer's training session, and spanned September 14 through October 22. By the end of October, 3,466 persons had been successfully surveyed, out of a total of 4,293 sampled persons, for a response rate of 81 percent.[29]

RAC knew that many citizens would not want to participate in a survey asking about personal, embarrassing, or illegal behaviors, and if they did participate, they might not answer the questions honestly. A high rate of refusal to participate in the study and a pattern of underreporting of marihuana use by those who did complete the interview could cause inaccurate results. So RAC used procedures to maximize cooperation and reporting accuracy of respondents. There was no explicit mention of the kinds of sensitive questions that respondents would be asked until the survey interview was well underway. At first, the interviewer read questions to the respondent, who answered aloud, and responses were entered by the interviewer (interviewer-administered). At the point in the interview where the drug use questions first appear, interviewers made strong promises of confidentiality, and allowed respondents to answer the sensitive questions using a self-administered questionnaire booklet. The interviewer handed the questionnaire to the respondent, and the respondent read the questions and entered the answers without telling the interviewer what they had entered or showing the completed answer booklet. This was done in a private setting away from other household members, if possible.

Three separate questionnaires were developed – two for adults and one for youths. Adults aged 18 and older received both an interviewer-administered and a self-administered questionnaire. The interviewer-administered portion came first, and obtained basic demographic information on respondents. It also contained questions on attitudes and opinions about drug use, including opinions about drug laws and penalties. Substance use questions were all self-administered. The total time to complete the adult interview was 40–45 minutes. Youths aged 12–17 received only a self-administered questionnaire designed specifically for their age group. It was shorter (about twenty-five minutes) and addressed a more limited range of demographic, attitudinal, and substance use questions.

The lead letter sent to all sample households prior to the interviewer's personal visit stated that if someone in the household was selected for the interview, they would be answering "questions about people's opinions on national and local issues." The title of the survey, "A Nationwide Study of Beliefs, Information and Experience" was intentionally vague so persons asked to participate would not become alarmed at the topic of the survey and decline to participate. Instructions established for interviewers when talking in person to household members after they received the lead letter were to avoid any mention of marihuana. The survey was about "social issues." And if a respondent asked, before the questionnaire was filled out, who sponsored the survey, interviewers were instructed to "say only that the survey is being done under a research grant." However, there was a limit imposed on this deception. The intentional camouflaging of the survey topic came to an abrupt end for adult respondents at question twelve of the self-administered questionnaire, roughly three-quarters of the way through the interview. The first questionnaire had no questions that would imply any personal culpability with illegal drugs. Thus, the interview was almost 75 percent complete by the time the respondent was asked directly if they had ever tried marihuana.

There was no Institutional Review Board or Office of Management and Budget to review and comment on the survey procedures.[30] The somewhat deceptive tactics employed in the initial National Survey were modified in later rounds of the survey, diminishing over time until the survey procedures included full disclosure of the survey's name, sponsor, purpose, and content in 2015.

The use of self-administered questionnaires within a face-to-face household survey to promote honest reporting of sensitive behaviors was not common at the time, and there was not much methodological research to justify the procedure. Intuitively it made sense that respondents would be more likely to report sensitive behaviors when an interviewer or household members would not know what their responses were. Methodological research done later proved this to be true.[31] But a drawback of self-administration is the problem of missing and inconsistent responses. Without interviewer control of the questioning or review of the respondent's completed answer sheet, some respondents will not follow instructions correctly, or may leave some answers blank. Additionally, responses may conflict with one another. For example, the response to a question asking "When was the most recent time you used marihuana?" could be "within the past week," and the response to a later question asking "During the past month, on how many different days did you use marihuana?" could be "No days." Obviously, one approach to minimize these kinds of errors is to make the layout and flow of the questionnaire simple and easy to follow. A thorough test of the questionnaire for ease of use prior to conducting the survey is essential. But inconsistencies and missing data are unavoidable. There are methods available to address the problem, post-data collection. Editing and imputation are commonly used approaches. Editing involves making decisions about how to "fix" the erroneous data by recoding answers according to a set of rules. Imputation requires filling in missing data through a statistical algorithm that may use responses to other questions, or may use information from other "similar" respondents who did answer the particular question. Methods for handling missing and inconsistent answers have changed several times over the course of the drug survey's history. For the 1971 survey, the data processing manual used by RAC coding staff simply instructs them to "Note interviews with any error code. Hold for review by Herb."[32] In later rounds of the survey, editing rules would become more standardized, complex, and fully documented.

After the sample data were edited, analysis weights were computed and attached to each individual respondent's data. These weights are numerical values that specify the relative contribution of each respondent's data to the overall estimates. The weight for a particular sample person is based primarily on their probability of being selected into the sample. The weights are designed so that when used in conjunction with the data collected on the survey, the resulting estimates represent the population that the survey is meant to cover. In the National Survey, population counts from the Census Bureau are incorporated into weights so that the weights for the entire sample will sum to the total population, and the weights for all respondents who report marihuana use sum to the estimated number of marihuana users in the country. A good way to think about survey weights is that if a respondent has a one in five thousand chance of being selected for the survey, then their analysis weight will be approximately 5,000 because they are representing approximately 5,000 persons in the population.[33] Weights must be used in making estimates from the survey data or in other types of analysis, not only because it is a sample, but also because the National Survey sample design has always assigned different selection probabilities to different population groups. For example, if the sampling plan requires youths to be sampled at twice the rate of adults, then weights for youths would be half of the weights for adults.[34] Adjustments to account for differential response rates across subpopulations and other factors are also incorporated into the computation of weights.

First Commission Report

A final report that included the analysis results was submitted to Dr. Susman on January 10, 1972. This information, along with reports of other commission activities and accomplishments, was published in March 1972 in compliance with the mandate to submit "a comprehensive report" to the president.[35] The 184-page main report was accompanied by a separate two-volume appendix (1,252 pages). The main report contained the survey results sprinkled throughout, supporting the various discussions on different aspects of the marihuana issue. The 1971 National Survey found widespread use of marihuana. An estimated 24 million Americans aged 12 and older had used marihuana at least once in their lifetime. Rates of use were highest among young people (see Figure 1.4).

Rates of lifetime use were 14 percent among youths age 12–17, 39 percent among young adults age 18–25, and 9 percent among adults age 26 and older. But the current extent of the problem was not as severe, as

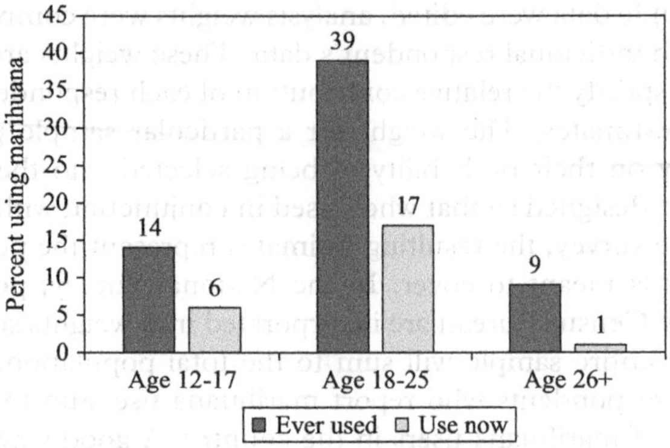

Figure 1.4 Marihuana use, by age, 1971.

fewer than half of people who had tried marihuana reported that they use "at the present time." The most common reason for quitting was simply a loss of interest in using marihuana. As indicated in the title, "Marihuana: A Signal of Misunderstanding," the report concluded that public fears about the threat of marihuana were unfounded. Seventy percent of adults and 56 percent of youths believed that marihuana makes people want to try stronger things like heroin; 56 percent of adults and 41 percent of youths believed that many crimes are committed by persons who are under the influence of marihuana; 48 percent of adults and 40 percent of youths believed that some people have died from using marihuana. The studies done by the commission showed that these public perceptions of the impact of marihuana were far from reality. Based on their research, the commission concluded that marihuana use at its current level did not pose a major threat to public health. They found no evidence of any health concerns associated with occasional use, and no direct link to criminal behavior or to use of other drugs.

Before the release of the report, President Nixon became aware of these findings. He summoned Shafer to the White House, and told him of his displeasure, warning him that the report needed to come out strongly against marihuana.[36] But Shafer refused to alter the thrust of the report. When the report was released in March 1972, it stressed the lack of evidence of any harmful health effects of marihuana. Furthermore, the report laid out several recommendations, including that while selling marihuana or possessing it with intent to sell should remain a felony, possession of marihuana for personal use should no longer be a federal offense. When asked about the report a few days after its release, President Nixon said "I oppose the legalization of marihuana, and that includes its sale, possession, and use. I do not believe you can have effective criminal justice based on the philosophy that something is half legal and half illegal. That is my position, despite what the commission has recommended."[37]

It's worth noting that despite the public misperceptions about the dangers posed by marihuana use, public support for Nixon's law enforcement approach was weak. When asked on the survey which course of action would be best to address the high rate of marihuana use among young people, only 37 percent of adults and 20 percent of youths favored "arrest, conviction, punishment." The preferred approach was "diagnosis, treatment, cure" among both adults (51 percent) and youths (48 percent).

Second National Survey

On July 11, 1972, RAC was awarded a second contract to conduct a similar drug survey. This time the survey content would be broadened to cover both marihuana and other drugs. In a shift towards transparency, the name chosen for round two was "Drug Experience, Attitudes, and Related Behavior Among Adolescents and Adults."

In planning and designing the second drug survey, Dr. Abelson continued to work closely with Dr. Susman. Several outside consultants also helped RAC on the study, including Professors Jack Elinson and Eric Josephson from Columbia University School of Public Health, Dr. Dean Manheimer, Director of the Institute for Research in Social Behavior in Berkeley, California, and Dr. Ira Cisin, Director of the Social Research Group at George Washington University. Cisin, who also had a faculty appointment with the University of California at Berkeley, had extensive experience in alcohol research. He had collaborated on studies with well-known alcohol researchers Don Cahalan and Robin Room. As discussed in Chapters 2 and 3, Abelson and Cisin would team up to conduct the survey for the next ten years, as it became established as a key data source for tracking progress in the war on drugs.

The methodology used for the second survey was nearly the same as in 1971. There were 200 interviewing locations, with a new sample of segments chosen within these locations. The within-household selection procedures remained the same as in 1971, except the objective for 1972 was to sample approximately equal numbers of the two age groups: 18–29 and 30 and over.

The questionnaires were the main departures from the 1971 methodology. There were five questionnaires: two for interviewer administration (adult version and a shorter youth version), and three self-administered questionnaires addressing: (1) cigarette and alcoholic beverage consumption; (2) experience and usage patterns for marihuana; and (3) experience and usage for heroin, LSD, cocaine, and inhalants. Each adult and youth respondent was asked to complete these three self-administered questionnaires at specified times during the interview. Total completion times for the entire interview process were 40–60 minutes for adults and approximately 30 minutes for youth.

Although the use of the word "drugs" in the second survey title may have represented a small move towards more disclosure of the survey's content and purpose, the opaqueness remained. From receipt of a lead letter through the start of the interview, the respondent was alerted to nothing more controversial than a request "for their opinions on a number of issues which are being talked about in the country these days." The interview began with a few benign questions. Then the respondent was offered the first self-administered questionnaire (questionnaire A), addressing cigarette and alcohol use. When completed, the interviewer instructed the respondent to place questionnaire A in the return envelope provided (to ensure confidentiality) and continued with more interviewer-administered questions. At that point, another innovation was introduced. A high-quality set of color images of about twenty-five prescription drugs was produced, including a printed list of ten common over-the-counter pills. These cards were handed to the respondent as a memory aid, to help them identify specific drugs that they had consumed. Once the interviewer resumed and asked another series of questions, self-administered questionnaire B on marihuana use was handed to the respondent to fill out on their own. Then there were five interviewer-administered questions on opinions of heroin and methadone. The third and last self-administered questionnaire (C) asked about use of heroin, LSD, cocaine, and inhalants. After a few more interviewer-administered demographic questions, the interview was complete. For youth (12–17) the procedure was similar, but considerably shorter.

Interviewer training was similar to that in 1971: one day sessions in nineteen cities throughout the country. Interviewing began on September 9 and concluded on October 15, 1972. There were 3,291 completed interviews. The interview response rate was 76 percent.

Rates of marihuana use were about the same in 1972 as they had been in 1971. An estimated 4.6 percent of adults and 4.8 percent of youths reported trying LSD or another hallucinogenic substance at least once in their lifetime. Adult and youth rates of cocaine use were 3.2 percent and 1.5 percent, respectively, and rates for heroin use were 1.3 percent and 0.6 percent. The 1972 survey found that a majority of the public mentioned drugs among the most serious problems in the country, about the same proportion as were concerned about the economy. The percent with concern about drugs was higher in 1972 than in 1971 (53 versus 44 percent).

The survey results were included in the commission's second report to the president and Congress, published in March 1973.[38] The commission made a series of recommendations, including the reaffirmation of its former findings. Decriminalization of marihuana possession was again recommended. The report concluded that there was either no relationship or a very trivial one between the use of marihuana and the use of other drugs. The report also found a weak connection between drug use and crime. The major drug dependence problems in the United States were said to be alcohol and heroin. Among its many specific recommendations were that Congress should create a single federal agency, separate from all federal departments and agencies, to administer and coordinate all drug policy at the federal level and to serve as the point of contact for state drug programs. This agency should develop and implement a drug research plan, and also "Maintain and monitor an ongoing collection of data necessary for present and prospective policy planning." No such comprehensive, independent agency was ever created, but drug data systems and research planning did soon emerge and flourish.

Notes

1 Kurlansky, *1968: The Year That Rocked the World.*

2 *New York Times*, 2/14/68

3 National Archives and Records Administration, "Statistical Information about Fatal Casualties."

4 Musto and Korsmeyer, *The Quest for Drug Control*, 60.

5 This spelling ("marihuana") is used throughout this chapter to coincide with the spelling used by the National Commission. Subsequent chapters will revert to the more modern spelling, "marijuana," although the survey questionnaire did not make the change until 1979.

6 *Time*, September 26, 1969.

7 Grim, *This Is Your Country*, 25.

8 Musto, "Opium, Cocaine and Marijuana," 42.

9 Morgan, *Drugs in America*, 29–32.

10 Grim, *This Is Your Country*, 24–26.

11 Morphine was first isolated from opium in about 1805. Heroin was synthesized from morphine in 1874, but it was not used much until 1898 when Bayer began selling it. Cocaine was first isolated from the coca plant in the late 1850s in Europe, and was commercially available in the United States by 1884.

12 Musto, "Opium, Cocaine and Marijuana," 45.

13 Robins, "Vietnam Veterans' Rapid Recovery."
14 Kuzmarov, *The Myth of the Addicted Army.*
15 Musto and Korsmeyer, *The Quest for Drug Control,* 48–53, 91–93, and 98–101.
16 Robins, "Vietnam Veterans' Rapid Recovery."
17 The percentages of heroin and marijuana initiates that served in Vietnam are based on retrospective reporting of age at first use and service in a combat zone during the Vietnam War era, based on pooled 2013–14 NSDUH data. SAMHSA added questions on military service in combat zones beginning with the 2013 survey.
18 Johnson et al., *Trends in the Incidence.*
19 Lee, *Smoke Signals,* 120.
20 National Commission on Marihuana and Drug Abuse, *Marihuana: A Signal,* Appendix Volume I: 289.
21 Musto and Korsmeyer, *The Quest for Drug Control,* 56–71.
22 Nixon, "Remarks on Signing."
23 Marihuana is not a narcotic.
24 Lee, *Smoke Signals,* 118–20.
25 Lee, *Smoke Signals,* 121.
26 Abelson, Cohen, Heaton, and Suder, *Public Attitudes Toward.* Like the first drug survey, the pornography survey was conducted to obtain data for a presidential commission (Commission on Obscenity and Pornography) to identify whether the presence of sexually explicit material in the United States was a social problem.
27 Miech et al., *Monitoring the Future.*
28 National Commission on Marihuana and Drug Abuse, *Marihuana: A Signal,* Appendix, Volume II: 855–1119.
29 Completed interviews and response rate were calculated from tables on pages 1037–40 in the first Commission report, Appendix Volume II, using method B described on pages 720–24 in the second Commission report appendix. Early National Survey reports did not contain the necessary data to calculate the household screening response rate, although the limited data indicates that it was near or above 90 percent every year until around 2008.
30 In the early 1970s, the federal government began requiring agencies to submit data collection plans to OMB for approval before surveys could be conducted. For the NSDUH, the first such submission occurred prior to the 1974 survey. Institutional Review Boards became prominent after the passage of the National Research Act of 1974.
31 Gfroerer and Kennet, "Collecting Survey Data."
32 National Commission on Marihuana and Drug Abuse, *Marihuana: A Signal,* Appendix, Volume II: 1,111.
33 In later years of the survey, materials used to convince potential respondents to agree to participate have included statements such as "Your participation is important because your responses represent 5,000 Americans."

34 Assigning higher probabilities of selection to certain populations (oversampling) is often done to ensure there is a big enough sample to analyze from that group separately.
35 National Commission on Marihuana and Drug Abuse, *Marihuana: A Signal.*
36 Lee, *Smoke Signals,* 121
37 Nixon, "The President's News Conference."
38 National Commission on Marihuana and Drug Abuse. *Drug Use in America.*

2 The Survey Continues, As Illicit Drug Use Peaks

While the Shafer Commission was doing its work, there was other activity in the war on drugs. In a June 17, 1971 executive order, the Nixon administration established the Special Action Office for Drug Abuse Prevention (SAODAP) in the Executive Office of the President. They also proposed legislation to formally establish the office. Their goal was for the White House to gain control of drug abuse programs in the Department of Health, Education, and Welfare (HEW). President Nixon did not trust HEW officials, saying that HEW personnel were "all on drugs."[1] SAODAP would coordinate the drug abuse "demand" activities (education, prevention, treatment and rehabilitation, and research) of the federal government. Nixon appointed Dr. Jerome Jaffe as SAODAP director – the first drug czar. Jaffe had been head of the drug abuse program of the Illinois Department of Mental Health. The Nixon administration liked his multimodality approach to treating heroin addicts that focused on methadone maintenance along with other techniques, with a recognition of the relationship between drug use and crime.[2]

After several months of debate, Congress and the White House finally agreed on a compromise bill authorizing SAODAP.[3] But the bill also set June 30, 1975 as the limit on the agency's existence. A major role for SAODAP was to lead the effort to expand federally funded treatment services, particularly methadone maintenance programs to treat heroin addiction. The bill also provided an additional $800 million for HEW special projects, and created the National Institute on Drug Abuse (NIDA), within the National Institute on Mental Health (NIMH), effective no later than December 1, 1974. NIDA would be responsible for developing and conducting comprehensive programs for the prevention and treatment of drug abuse and the rehabilitation of drug abusers. NIDA would take over much of SAODAP's work once SAODAP expired. At that point, direct management of these activities would shift from the White House to HEW.[4]

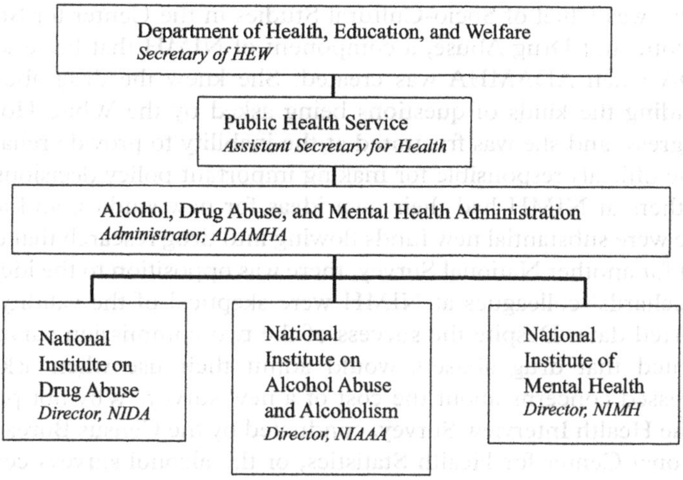

Figure 2.1 Creation of ADAMHA, 1973.

Well before the December 1, 1974 deadline, HEW created NIDA, but not as a sub-agency within NIMH.[5] The drug abuse institute would be part of a new agency, the Alcohol, Drug Abuse, and Mental Health Administration (ADAMHA). NIMH and the National Institute on Alcohol Abuse and Alcoholism (NIAAA) would also be placed in ADAMHA. As diagrammed in Figure 2.1, under this reorganization, NIDA and NIAAA became separate and equal to NIMH, instead of under NIMH control.

NIDA and NIAAA would be able to develop their own approaches and direct their research programs more towards their specific needs and issues. The substance abuse agencies would not be limited by the view in the mental health community that substance abuse was a symptom of mental illness. Also, the opposition to methadone maintenance among NIMH leaders would not hinder NIDA's support and research for this mode of treatment.[6] Dr. Robert DuPont was named NIDA's first director. At the time, he was also holding the position of SAODAP director (Jaffe had resigned in July 1973) as that agency wound down.[7]

1974 Survey: Public Experience with Psychoactive Substances

In late 1972 or early 1973, SAODAP issued a call for new ideas concerning substance abuse programs. Dr. Louise Richards had an idea: continue the National Survey. She knew that SAODAP supported it.

Louise was Chief of Socio-Cultural Studies in the Center for Studies of Narcotic and Drug Abuse, a component of NIMH that became part of NIDA when ADAMHA was created. She knew the drug abuse field, including the kinds of questions being asked by the White House and Congress, and she was frustrated at the inability to provide reliable data to the officials responsible for making important policy decisions.

Others at NIMH had their own ideas for new studies, and although there were substantial new funds flowing into drug research that could be used for another National Survey, there was opposition to the idea. Some of Richards' colleagues at NIMH were skeptical of the validity of self-reported data. Despite the success of the two commission surveys, they doubted that drug abusers would admit their use when asked, and expressed concerns about the cost of a new survey. Why not piggyback on the Health Interview Survey, conducted by the Census Bureau for the National Center for Health Statistics, or the alcohol surveys conducted by the Alcohol Institute? When asked, these agencies showed little interest. Richards recounts, "Those agencies were skittish about adding drugs to their surveys," feeling that this was a very sensitive area.[8]

Had the piggyback concept materialized it would not have endured. Access to interview time would be inadequate to satisfy the future demands for drug abuse information, not to mention the loss of managerial control over the survey methodology. Getting the project off the ground in the context of the federal bureaucracy would not be as easy as it had been for the commission. There were federal procurement regulations and time-consuming clearances. But Louise had been operating in the federal government world for quite a while and was used to the bureaucracy. Her knowledge, social skills, and leadership ultimately won the day. She consulted with colleagues to refine the survey plan, adding a pre-survey validation study to the statement of work. If the validity study showed that drug users would lie about their use, the survey would not be carried out. Finally, with all detractors either disarmed or pacified, and with the approval of the appropriate SAODAP official,[9] Louise Richards proceeded to ask the NIMH contract management office to issue a request for proposals to conduct the 1974 survey.[10]

Proposals were received in late May 1973, and an evaluation committee met to rate the four qualified proposals. The top-rated proposal was from George Washington University's (GWU) Social Research Group, led by Dr. Ira Cisin, with Response Analysis Corporation (RAC) as a sub-contractor responsible for data collection. Richards was familiar with the commission's surveys and was convinced that Abelson and Cisin "were very able scientists, not simply market researchers," and felt that they would be an excellent choice for bringing about the survey's revival.

The contract was awarded on June 28, 1973, with Richards named as the project officer.[11] The original contract was signed under the letterhead of the NIMH, but by the initial planning meeting in September it was under the jurisdiction of the newly created NIDA.

The statement of work in the contract was amazingly brief by current standards, as was the time frame for completing the work. There were only two pages of specifications. A small validity study was to be performed prior to the main survey, with final results to be applied to questionnaire development and overall methodology to maximize the accuracy of responses. A sample size of 2,000 was specified for the main survey, with a sample design similar to the 1972 survey. All of the work was to be completed in one year, with the contract ending on June 27, 1974. A response rate of 75 percent was specified for each major age and race/ethnic category, and region. Given the scope of the project, the sequential aspect (the validity study needed to be completed before the design of the main survey could be finalized), and most importantly, the requirements for clearance of the study design by the Office of Management and Budget, the plan was doomed from the start. Richards underestimated the realities of executing a national survey, and overestimated the capabilities of the contractor. The government clearance and approval process took nine months. There were problems encountered in designing and conducting the validity study. Ira Cisin had extensive experience validating survey responses on alcohol, but not drugs. The study plan involved extracting patient records from drug abuse treatment clinics, and conducting interviews with these patients. On June 21, 1974 Cisin wrote a letter to Louise Richards explaining that the design that "had been used with remarkable success in our validity study on drinking practices some years ago was failing miserably" when applied to the subject of drug abuse. There were unanticipated problems with gaining clinic cooperation, concerns about confidentiality, incomplete and inaccurate records at the clinics, and difficulties in locating former drug clients. According to Cisin, drug clients give the clinics fictitious names and addresses; clinics were staffed by inexperienced personnel and make many clerical errors; and many drug clients have fluid living arrangements. Given these problems and the delay in getting clearance, Cisin recommended "that we go ahead with the planning of the national household study on schedule, instead of postponing it until the completion of the validity study ... our qualitative content analysis of the work completed on the validity study so far leads me to conclude that there is no doubt about the ability of the data collection procedure to yield reasonably valid results." NIDA agreed, and the main survey work proceeded.

A modification to the contract extended the period of performance to March 31, 1975, increased the sample size by 2,000 respondents, added items to the questionnaire, and brought in a group of consultants to assist the contractor. Interviewing for the survey began in mid-November 1974, not an ideal time of year, due to inclement weather and holidays. Not surprisingly, due to difficulties in obtaining interviews, data collection was not completed until March 13, 1975. The contract had to be extended again to allow time for writing a report on the findings.

The 600 RAC sample locations from which the 1971 and 1972 National Survey samples were drawn were used as the initial stage of the 1974 sample. A new subsample of 400 locations was drawn for use in 1974, and new sample segments were selected within these areas. Field procedures were similar to those employed in the 1972 survey. A total of 4,023 persons were interviewed and an interview response rate of 83 percent was achieved.

There were some methodological differences between the 1974 survey and the earlier surveys. RAC employed an interviewer bonus program, an attempt to achieve high response rates through monetary incentives given to the interviewers. As in the earlier surveys, virtually all survey personnel were kept in the dark with respect to the sponsor of the survey. Field supervisors told interviewers to tell respondents it was the George Washington University. Even the supervisors were told only that HEW and GWU were the sponsors, but that it was GWU "who is paying us." There was no mention of the National Institute on Drug Abuse. The lead letter described the study as "an important public opinion survey," and the title on the questionnaires introduced a generic "Current Trends" survey. However, the interviewer did inform the respondent early in the interview that most of the questions "are about alcohol, tobacco, and other drugs." On the adult questionnaire, this occurred after a few opinion questions, just before the interviewer asked "During the past month, have you smoked any cigarettes?" In the youth interview, the disclosure occurred at the very beginning of the interview, since the smoking question came first.

Spanish language lead letters and questionnaires were used for the first time in 1974. When an interviewer encountered a Spanish-speaking household they had the assignment transferred to a Spanish-speaking interviewer.

The interview format and instruments for 1974 were somewhat different than in the previous surveys. The adult interview was forty to sixty minutes, and the youth interview was about thirty minutes. There was an interviewer-administered and interviewer-marked questionnaire that included questions on the use of cigarettes, coffee, tea, alcohol, and medical and nonmedical use of prescription drugs. Youth respondents were given a shorter version covering these same substances. Following the interviewer-administered component, respondents were first asked to complete a self-administered questionnaire on marijuana use. Upon completion, the respondent placed it in the large return envelope. Next came an interviewer-administered questionnaire associated with seven different colored answer sheets sequentially handed to the respondent for the purpose of answering confidentially. Each answer sheet addressed a specific type of drug (hash, inhalants, cocaine, hallucinogens, heroin, methadone, and opium). There were no questions on these answer sheets. Interviewers asked the questions and respondents wrote in their answers. Interviewers were to make every effort to ensure the respondent was comfortable about the privacy of the environment. Each colored answer sheet was placed in the return envelope after it was completed. The survey continued with an interviewer-administered and marked procedure, consisting of a few minutes of opinion, health, and demographic questions. The last question was on family income (not asked of youths).

A new feature of the marijuana self-administered questionnaire was the removal of all skip patterns. Each respondent was supposed to answer every question, instead of skipping over questions that were not relevant. For example, if the respondent answered in the beginning that they had never used marijuana, they still had to answer subsequent questions about recency and frequency of marijuana use. Each question included a response category for nonusers to select ("Never Used"). This approach was a change from the prior two surveys, in which respondents were sometimes instructed to skip to a later question if they gave a particular response to a question. For example, in 1972 if a respondent replied on question C17 that they had never tried LSD, the questionnaire instructed them to skip to question C22, avoiding the questions asking about their experiences with LSD. The absence of skips would be a hallmark of the survey for the next twenty-five years. There were several reasons for using this procedure, despite the potential for it to seem awkward and annoying to some respondents. First, it would reduce erroneous skipping. Second, it would assure respondents that their answers were private and would not be revealed or surmised by the interviewer or by other household members nearby. The reasoning was that every respondent, regardless of whether or not they had used a drug, would take about the same amount of time to complete each self-administered questionnaire sheet. With skips, nonusers would generally finish filling out the answers more quickly than drug users. Finally, there was concern that with a series of self-administered questionnaires in which a response indicating no drug use allows the respondent to skip out of the remainder of the questions, respondents would figure out after one or two of these sheets that they could speed up the interview by denying use of each drug. In other words, skip patterns that result in a shorter interview for nonusers could motivate some drug-using respondents to deny their drug use. An experiment done in 1992 found that this does occur.[12] One downside to removing all skips from the self-administered questionnaires is the substantial post-data collection processing that becomes necessary due to the frequent blanks and inconsistent responses occurring on the forms.

The results of the 1974 National Survey were submitted by the contractor to the project officer in much the same way they had been submitted for the commission surveys. There were four parts. Part 1 summarized the results, with narrative and tables, including a short description of the survey methods. It was referred to as the "Main Findings" report. Part 2 was a large set of detailed tabulations. Part 3 was a detailed report of the methods and procedures of the survey. Part 4 was a magnetic data tape containing all of the data. This general format was maintained for future rounds of the survey, although additional reports were added for some of the surveys. In particular, the "Population Estimates" became a regular report. It contained estimates of the number of users of each substance, in contrast with the Main Findings reports, which presented percentages. Another report that was occasionally published was a "Highlights" report which gave a brief summary of the most important results from the survey, and was used as a handout in some of the press conferences releasing the latest findings.

1976 Survey: Nonmedical Use of Psychoactive Substances

Before the 1974 National Survey contract ended, NIDA issued another contract modification. "The Director of NIDA finds it imperative to report annual rather than biennial statistics to HEW and Congress. This study must be conducted in the fall of 1975 to be compatible with the results of the 1974 and 1972 studies," the contract modification stated. Of course, the 1974 survey had not actually been conducted in the fall. Project officer Louise Richards was opposed to doing the survey every year. She later recalled that "the contractors also thought it was too hard to turn around in one year, because they knew that it had taken longer than everyone had expected for the first one to be completed." Nevertheless, the order from NIDA Director DuPont to conduct the interim 1975 survey in order to provide data sooner had to be carried out. Although it meant that a lot of work needed to be done quickly, DuPont's directive showed that the survey was perceived to be of great value, and had rapidly attained a high profile.

The only way to satisfy the new requirement was to have the next survey done by the same firm, under the same contract. There would not be time to go through the entire procurement process, nor risk having a new, inexperienced firm responsible for starting a new survey from scratch. The GWU-RAC contract was extended again and additional funds were added to cover the cost of the 1975 National Survey, which as it turned out was conducted from January through May 1976. It was later referred to as the 1975/76 survey or simply the 1976 survey.

The technical and administrative demands of the survey became a burden for project officer Louise Richards when she became head of the newly formed Psychosocial Research Branch, within the Division of Research. "I never got to do all the things that I dreamed of doing or wanted to do with the survey data … it always seemed there were so many other pressures," she later recalled. She decided to delegate some of her survey responsibilities to her staff, allowing her an opportunity to begin generating some secondary analysis. In mid-1975, Dr. Joan D. Rittenhouse joined Richards' branch as a research psychologist. The primary motivation for Rittenhouse accepting the position was her perception that the survey, "was a goldmine for seminal research," and that she would become the project officer of an important government research tool. She began as the alternate project officer, but was the de facto head of the survey from the beginning of her tenure at NIDA. By the beginning of data collection for the 1976 survey she had attained official project officer status. Under her supervision, the project would have more emphasis on secondary analysis. The contractor was encouraged to investigate the correlates of drug use and abuse, the consequences of drug use, and other research associated with the National Survey data. Rittenhouse would later recall, "Louise had vision." Joan was intent on continuing the legacy during her watch.

With the unreasonable requirement imposed in mid-1975 to conduct the fourth National Survey in the fall of the same year, only a near clone of the 1974 design would foster a successful survey. Nevertheless, a few changes were made:

- The interviewer bonus program was discontinued.
- The lead letter still referred to "a public opinion survey on issues of national importance," with no mention of the topic. But there was a new follow-up letter sent to some sampled households that interviewers had difficulty contacting. This letter informed the resident that

"The purpose of the study is to better understand how people feel about alcohol, tobacco, and other drugs."

- All respondents were informed at the beginning of the interview that most of the questions were about alcohol, tobacco, and other drugs.
- The opinion questions at the beginning of the adult interview were dropped. Thus, respondents no longer were gradually introduced to the substance use questions; instead the cigarette questions were first for adults as well as youths.
- The number of questions on prescription drugs was reduced.
- Questions addressing frequency of use and year of first use were made more consistent across drug categories.
- Adult and youth versions of the instruments were made almost identical.

The statement of work called for a sample size of 3,500 individuals 12 years and older. The field effort required about the same number of interviewers (334) as in 1974, but training costs were considerably less, since only newly hired interviewers (about a quarter of the total) were given formal one-day sessions. The experienced interviewers were given only a quiz and a couple of practice interviews to do at home. Although the contract specified data collection at the same time of year as in the previous survey (November through March), interviewing did not begin until January 1976. Data collection concluded in May, with 3,576 completed interviews and a response rate of 84 percent.

1977 and 1979 National Surveys on Drug Abuse (NSDA)

With the "interim" survey conducted in early 1976, and the scheduled 1977 survey preparation underway, the National Survey was, for this brief period, conducted annually. However, the original idea of conducting the survey every two years prevailed, as the next survey was planned for 1979. Separate contracts for the 1977 and 1979 surveys were put out for competitive bids, and each time, the independent review process resulted in the GWU-RAC team winning. Joan Rittenhouse, under the supervision of Louise Richards, remained as project officer for these rounds of the survey, which was officially named the National Survey on Drug Abuse (NSDA) beginning with the 1977 survey.

The 1977 NSDA design was similar to the 1976 design. The preface in the Main Findings report from the 1977 survey states "Dr. Joan Dunne Rittenhouse, Project Officer for the sponsoring NIDA, has contributed significantly to the shaping and execution of this study; beginning with the specifications for a high quality research design, she has participated in every step of the strategy planning and assumed responsibility for all final decisions."

The sample of segments was selected from 400 interviewing locations, which had been sub-selected from the original 600 locations in the standing RAC sample. Interviewing took place from March through July 1977. A total of 4,594 persons were interviewed, and the interview response rate was 85 percent. The questionnaire was much the same as in 1976, although some prescription drug questions were dropped, and the remaining prescription drug questions were administered to half of the sample. A new set of questions asking about heroin use by friends was added.

The 1979 NSDA featured a supplemental sample in rural areas. The basic sample was similar to the design used in 1977, but an additional 100 secondary sampling units were selected in rural areas, defined as nonmetropolitan communities with a population of less than 25,000. Data were obtained from 7,224 respondents during August 1979 through January 1980. The response rate was 86 percent. There were changes in the questionnaire. Coffee and tea consumption questions were dropped in 1979. Questions on medical use of prescription drugs were dropped. Cigarette and nonmedical prescription drug questions remained in the interviewer-administered part of the interview, but the alcohol questions were moved to a self-administered questionnaire. As a result, the alcohol module became the first self-administered questionnaire encountered by respondents, followed by the marijuana module (the first self-administered module in 1977). In addition, the marijuana module was merged with the hashish module, and became the marijuana and hashish module, because they are basically the same substance (cannabis). Another change to the self-administered questionnaires was the inclusion of the full wording of each question on the answer sheets. For example, in 1977, the interviewer read the question "When was the most recent time that you used cocaine?" aloud, and the respondent selected an answer category under the abbreviated "Most recent time?" shown on the answer sheet. In 1979, the full wording of the question, identical to what the interviewer read, appeared above the response options on the answer sheet. At the beginning of the third answer sheet (inhalants), the respondent was told that if they preferred, the interviewer could stop reading all of the questions aloud, and the respondent could complete all of the remaining answer sheets on their own. There were other wording changes and shifts in question ordering between 1977 and 1979. Taken as a whole, the changes may have affected comparability between the 1979 survey and the prior surveys. The 1979 Main Findings report underplayed the potential effects of these changes, containing only one caution regarding the changes and their impact on comparability. The change in mode of administration (from interviewer to self-administered) for alcohol is noted, warning that the data are not comparable.[13] The report explains that "The new design was implemented to (1) provide respondent training on the answer sheet procedure prior to its use for illicit substances, and (2) provide the same conditions of privacy for this drug as for the illicit drugs, thus encouraging full disclosure." Four pages later, a table showing estimates of current drinking appears. The rates for youths are 34 percent in 1974, 32 percent in 1976, and 31 percent in 1977. The 1979 estimate is 37 percent. Similarly, 6 percent jumps in the 1979 drinking rates following several years of steady rates are seen for young adults and for older adults. So did drinking rates really go up in 1979? Or was the 6 point increase an artifact of the change in the way the alcohol questions were asked? It's impossible to tell for sure. More troubling is the possibility that the changes to the questionnaire could have affected reporting of marijuana and other drugs. It would have been possible to assess the effect of the questionnaire changes by retaining the old questionnaire for a random half of the sample. But that would have been difficult to implement, because interviewers would have had to be trained to administer the survey both ways, and select the right version for each interview so the assignment of questionnaires was purely random. Also, splitting the sample in this way means that only half of the sample can be used for some analyses and estimates. The choice for the survey team in designing the 1979 NSDA was between improving the quality of the data, or maintaining strict comparability with an inferior methodology so that trends over time could be measured. It's a common dilemma for ongoing surveys, and it was faced many times throughout the National Survey's history.

Estimating the Prevalence of Heroin Use

President Nixon's war on drugs was partly in response to the surge in heroin use and addiction in the US population and among US troops in Vietnam. Reliable estimates of the size and characteristics of the heroin problem were needed to determine the best approaches to combat the problem and to track progress after programs were implemented. But there was little data available. Even before the first commission survey, there were attempts to develop estimates of the size of the heroin addict population. One method was the capture-recapture procedure.[14] This method has been used to estimate the number of fish in a lake or wildlife in a large area, by capturing a sample, tagging them, releasing them, allowing them to mix, and then recapturing a new sample and counting how many of the new sample are tagged. Simple projections can easily be made from this sampling and tagging process, but the accuracy of the estimates depends on some basic assumptions about the nature of the samples and populations studied. It's important for each of the samples to be random, and for the tagged individuals to fully disperse across the sampled area before the second sample is drawn. But the application is questionable, because heroin addicts do not behave like fish in a lake or deer in the woods. They don't enter treatment or get arrested (tagged) randomly, and they don't disperse fully into the population after release. Nevertheless, capture-recapture was one of several methods that researchers have used to estimate heroin use or addiction. From 1968 to 1974, various studies estimated the number of heroin addicts in the United States, with numbers ranging from 250,000 to more than 600,000.[15] But even during the height of discussions about legislation, policies and funding to address the heroin problem, Drug Czar Jerome Jaffe in 1972 cautioned that the available estimates were crude and "the fewer times we mention them, the better."[16]

The National Survey began including questions on heroin use in 1972. But the survey estimates had limited usefulness during the 1970s and 1980s because of the small sample sizes, coupled with the relatively low rate of heroin use. With past year use rates consistently well below a half of a percent of the population, the sample would only capture a handful of users each year – too few to make reliable estimates. The estimated lifetime prevalence of heroin use (about 1 percent of the population during the 1970s) became the principal indicator used in the survey reports and in tracking trends.

Sample size limitations can be addressed by selecting larger samples or by pooling multiple years of data during analysis. However, there were other concerns about the accuracy of heroin use estimates. It was widely believed that a high proportion of heroin users, addicts in particular, would not be captured in a household survey sample (coverage error). They are referred to as a "hidden" or "hard-to-reach" population, requiring specialized sampling and data collection methods for study. Also, many of the heroin users that complete the interview would be reluctant to admit to this highly stigmatized behavior (underreporting). Because of these concerns, heroin estimates from the National Survey are always used with caution, with a description of these caveats accompanying discussions of results. As is the case with other drug use estimates that may have some level of underestimation associated with them, there is some confidence in the accuracy of trend measurement, based on the assumption that the underestimation is constant from year to year.

Prior to the 1977 National Survey, project officer Joan Rittenhouse explored an interesting new approach for estimating heroin use from the survey. The nominative technique was developed by Dr. Monroe Sirken, a leading statistician at the NCHS. It was an adaptation of network sampling, a method that Sirken had developed for use in estimating rare diseases and health conditions from a general population survey.[17] To estimate heroin use, a subsample of NSDA respondents were given a set of interviewer-administered questions asking how many close friends they have, how many of them have used heroin, characteristics of the heroin users, and how many of each heroin user's other close friends knew that they had used heroin. From these responses, a national estimate of the number of heroin users was derived. Sirken's method essentially provided a way to cast a wider net in order to find more heroin users. In addition, by reporting about friends' heroin use (without identifying them) instead of their own, the respondents' fear of revealing sensitive, embarrassing information was diminished. The nominative questions were included in the National Surveys of 1977, 1979, and 1982. The nominative method estimated 1,880,000 past year heroin users in 1982, several times larger than the standard estimates based on direct questioning in the self-administered questionnaires.[18] The method did produce some unusual results, such as dramatic, unrealistic shifts in prevalence rates between 1979 and 1982, and was discontinued in the next (1985) survey. The questionnaire continued to include the direct questions on heroin use, and estimates have been published every year, along with appropriate caveats about the likely undercoverage and underreporting. In later years, when the sample size of the survey was greatly increased, estimates were more widely used, but still with caution by those who understood their weaknesses.

The Drug War's Impact on Drug Abuse

In September 1973, ADAMHA and NIDA were created, and the Shafer Commission ended its work. That same month, there was another key milestone in the war on drugs. In a speech at the White House to a group of criminal justice representatives from twenty-three cities and counties, President Nixon proclaimed success, and concluded his remarks with an attempt at humor:

> We have turned the corner on drug addiction in the United States–haven't solved the problem, because we have a long way to go. There is a long road after turning that corner before we get to our goal of getting it really under control, but we have

turned the corner. The numbers, the statistics, are beginning to be better ... Let me say, as a result of the trip you have taken to Washington, we know that many young people in this country will not be taking unhappy trips to other places through other means in the future.[19]

Evidence supporting the claim of success in the drug war was tenuous, and primarily addressed heroin addiction. Results from the first two National Surveys did not indicate any decline in drug abuse. The administration claimed that the heroin supply was down, price was up, seizures and arrests were up, heroin-related crime was down, treatment was more widely available, and there was a decrease in heroin initiation. The accuracy of these data and what they implied were questioned. Later analyses would contradict the claim that heroin use had been reduced, showing that the highest rate of heroin initiation occurred in 1973. It was not until 1978 that a substantial reduction in initiation occurred.[20] Nevertheless, it was evident by 1973 that public interest in the drug war was waning.[21] Also, with Nixon winning reelection in 1972, the administration no longer needed to address drug abuse as a political issue. They were preoccupied with Watergate and bringing the Vietnam War to an end.

After Nixon resigned in August 1974, President Ford and the Congress continued to downplay the drug war and tone down the war rhetoric. The president wanted to restore more authority and responsibility to cabinet departments and federal agencies, in the wake of the Watergate scandal and the associated abuses of power by the Nixon White House. Ford relied heavily on a white paper prepared by his Domestic Council, released October 14, 1975, to define his drug policy.[22] The report stated:

> We should stop raising unrealistic expectations of total elimination of drug abuse from our society. At the same time, we should in no way signal tacit acceptance of drug abuse or a lessened commitment to continue aggressive efforts aimed at eliminating it entirely. The sobering fact is that some members of any society will seek escape from the stresses of life through drug use. Prevention, education, treatment, and rehabilitation will curtail their number, but will not eliminate drug use entirely. As long as there is demand, criminal drug traffickers will make some supply available, provided that the potential profits outweigh the risks of detection and punishment. Vigorous supply reduction efforts will reduce, but not eliminate, supply. And reduction in the supply of one drug may only cause abuse prone individuals to turn to another substance. All of this indicates that, regrettably, we will probably always have a drug problem of some proportion. Therefore we must be prepared to continue our efforts and our commitment indefinitely, in order to contain the problem at a minimal level, and in order to minimize the adverse social costs of drug abuse.

Despite President Ford's desire to shift responsibilities from the White House to departments and agencies, in March 1976, Congress passed legislation creating the White House Office of Drug Abuse Policy (ODAP). Ford signed the legislation, but only because the bill also contained important funding authorizations for drug treatment and prevention. When he signed the bill, Ford said:

I have voiced strong opposition to the reestablishment of a special office for drug abuse in the White House. I believe that such an office would be duplicative and unnecessary and that it would detract from strong Cabinet management of the Federal drug abuse program. Therefore, while I am signing this bill because of the need for Federal funds for drug abuse prevention and treatment, I do not intend to seek appropriations for the new Office of Drug Abuse Policy created by the bill.

SAODAP, which came to an end on June 30, 1975, had responsibility for oversight of only demand-side activities. Although ODAP was small and had limited authority, the scope of its coordination responsibilities included both demand and supply-side programs. But budget cutting during the Carter administration prevented ODAP from lasting beyond 1978. Carter continued the tolerant attitude towards marijuana that had been articulated by the Domestic Council White Paper, and even proposed decriminalization, as did some members of Congress, citing polls showing public support for civil fines rather than imprisonment.[23] Between 1973 and 1977, eleven states decriminalized marijuana possession. Carter's chief drug policy advisor, Dr. Peter Bourne, advocated decriminalization of marijuana, and advised Carter in 1977 that cocaine had no serious medical consequences and coca eradication efforts in Bolivia should be curtailed.[24] Bourne was forced to resign in 1978 after he was caught writing a falsified prescription for a member of his staff, and there was a report that he had used cocaine at a party hosted by the National Organization for the Reform of Marijuana Laws, a group advocating marijuana legalization.[25] After Bourne departed, the White House tone shifted away from tolerance and more towards concern about the increasing marijuana and cocaine use among young people. The shift was partly in response to a growing grassroots parents movement, which formed out of frustration with the laid back attitudes and policies of politicians and health professionals at a time when there was a growing crisis of drug use by youth, evident in data from NSDA and other surveys. The public's concern about the drug problem had diminished throughout the 1970s. Economic problems such as inflation and unemployment were the greatest concerns. However, as the decade and the Carter Administration came to a close, and the rates of use among

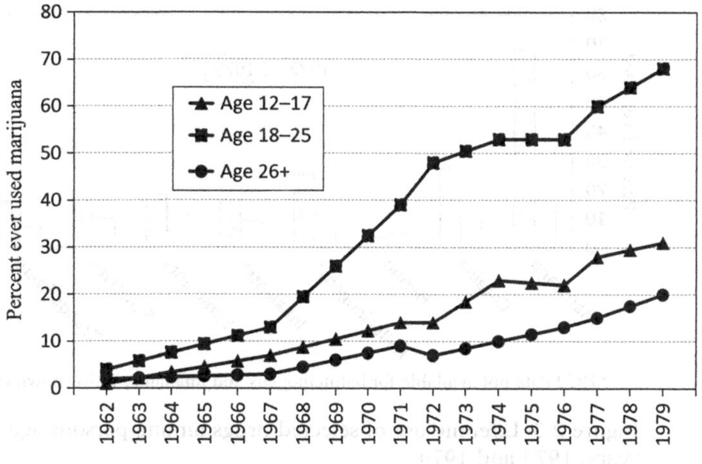

Figure 2.2 Lifetime use of marijuana, by age and year.

youth reached their highest levels ever, the tide seemed to be turning towards a more prominent anti-drug use attitude. It would turn into a tidal wave during the next administration.

So what effect did Nixon's drug war, first announced in 1969, have on drug use over the next decade? With five comprehensive national drug surveys conducted from 1972 to 1979, the federal government had the capability to assess trends over time in some detail. Although there were comparability problems caused by the methodological changes to the survey, the effects of many of those changes may have been small, relative to the magnitude of the sampling errors during those years. In other words, with the small sample sizes in those early surveys, only very large differences in prevalence rates would be statistically significant, and the size of those differences would be much larger than the methodological effects.

The Department of Health and Human Services (HHS)[26] held a press conference on June 20, 1980 to release the 1979 National Survey results. A special retrospective incidence report based on the 1977 data was also presented.[27] The new data, shown in Figure 2.2, quantified the rise of marijuana use among young people during the 1960s, a major impetus for the drug war.

But the surveys showed that substantial increases continued after Nixon's announcement of success in 1973. In 1962, only 1 percent of youths aged 12–17 had ever used marijuana, and in 1972 the rate was 14 percent. By 1979 it reached 31 percent. Among young adults age 18–25,

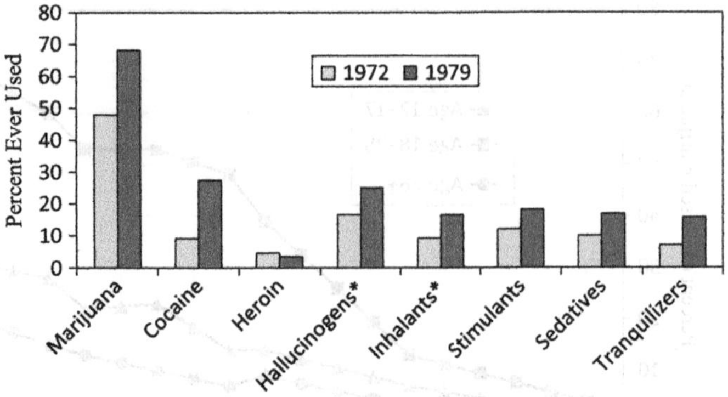

*1972 data not available for hallucinogens and inhalants; 1974 shown instead.

Figure 2.3 Lifetime use of selected drugs among persons age 18–25 years, 1972 and 1979.

rates were 4 percent in 1962, 48 percent in 1972, and 68 percent in 1979. The 1979 survey estimated that 55 million Americans had used marijuana in their lifetime, more than double the number estimated from the 1971 survey. Furthermore, 23 million were current (past month) users in 1979. There were also substantial increases in lifetime use of cocaine between 1972 and 1979 (see Figure 2.3), although at lower levels than marijuana, and mainly among the young adult age group (from 9 percent in 1972 to 28 percent in 1979).

The release of these alarming new survey results did not include any assessments of progress in the war on drugs. There was no claim of success or failure of any program, and no explanation of the reasons for the increases. NIDA and HHS officials seemed to recognize that the NSDA was not designed to determine the causes of drug use, or to evaluate the effectiveness of particular prevention or treatment approaches for reducing the impact of drug use. The survey was simply a tool to estimate the size of the drug abuse problem, variation across population subgroups, and changes over time. The survey could report on the correlates of drug abuse – factors associated with drug abuse, and *potential* causes for use patterns – and some of the consequences of use. In the 1980 press release, HHS Secretary Patricia Harris said "These two reports show that the deep concerns of the American people in general, and parents in particular, about the rapid rise in illicit drug use over the past few years are well founded." Harris went on to say that she was directing NIDA to "intensify its ongoing efforts on the prevention of drug abuse."

It is likely that this press release, issued less than five months before the 1980 presidential election, was carefully worded and possibly timed with the election in mind. In particular, the acknowledgement of the parents movement and the promise that the federal government would do more to prevent youth drug abuse may have been emphasized in order to blunt what otherwise was very bad news for the nation and for an incumbent presidential candidate.

Notes

1 Musto and Korsmeyer, *The Quest for Drug Control*, 90.
2 Musto and Korsmeyer, 77.
3 The Drug Abuse Office and Treatment Act of 1972 (Public Law 92–255, March 21, 1972).
4 Rettig and Yarmolinsky, *Federal Regulation of Methadone Treatment*.
5 NIDA was formed primarily by combining the NIMH Division of Narcotic Addiction and Drug Abuse with components of SAODAP. The reorganization took place in September 1973. Additional staff and work were transferred from SAODAP to NIDA in 1975, when SAODAP came to an end.
6 DuPont, "Present at the Creation," 82–7.
7 DuPont, 84–5.
8 Louise Richards, personal interview, August 4, 1981.
9 John Ball, SAODAP, memorandum to Louise Richards, April 20, 1973.
10 Louise Richards, memorandum to Chief, Contracts Management, April 25,1973, requesting a request for proposals (No. NIMH-NDA-73–197), titled "Nationwide Drug Abuse Survey."
11 "Project Officer" is the Federal Government's term referring to the technical representative of the government, responsible for monitoring a contract.
12 Gfroerer, Lessler, and Parsley, "Studies of Nonresponse," 273–295.
13 Fishburne, Abelson, and Cisin. *National Survey on Drug Abuse*, 89.
14 Woodward, Bonett, and Brecht. "Estimating the Size," 158–171.
15 Musto and Korsmeyer, 101–105
16 Markham, "The Heroin Addict Numbers Game." *New York Times*, June 4, 1972.
17 Sirken, "Network Surveys," 31–32.
18 Miller, "The Nominative Technique," 104–124.
19 Nixon, "Remarks at the First," 8.
20 Johnson et al., "Trends in the Incidence."
21 Musto and Korsmeyer, 144–145.The Gallup and Harris Polls showed that public concern about the drug problem peaked in 1972 and 1973, and declined sharply in 1974. The high cost of living, inflation, and unemployment were the top concerns of the public during 1974–1976.
22 Domestic Council Drug Abuse Task Force, *White Paper on Drug Abuse*.
23 Musto and Korsmeyer, 192–193.
24 Musto and Korsmeyer, 186–215.
25 Shaffer, Ronald. "Cocaine-Sniffing Incident," *Washington Post*, July 21, 1978.

26 The United States Department of Health, Education, and Welfare (HEW) was a cabinet-level department of the United States government from 1953 until 1979. In 1979, a separate Department of Education was created from this department, and HEW was renamed as the Department of Health and Human Services (HHS).
27 Miller and Cisin, *Highlights*, 13–16.

Wrap-Up for Chapters 1 and 2

Looking back on the genesis of the survey and its first decade, it can be seen that many of the factors that affected the survey throughout its history were evident from the start.

Social and political influences indirectly led to the launching of the first national drug survey in 1971. The Vietnam War fueled a rise in heroin addiction in the United States. Marijuana and other illegal drug use also seemed to be increasing among youth and young adults, and President Nixon perceived anti-war protesters and the broader youth counterculture not only as his political enemies, but as the groups that were driving this increase. These factors prompted Nixon to announce a War on Drugs and form a national commission to study drug abuse. The commission then decided to conduct the survey, recognizing the need for national data on the growing drug problem.

However, when the commission decided to conduct the survey, the scope of the nation's drug problem was far from its peak. Ten years later, rates of use were much greater, especially among younger teens. So even if the commission had not been formed and the two surveys were not conducted in 1971 and 1972, it was inevitable that the federal government would have eventually decided to conduct a national drug survey. During the 1970s, the drug problem was getting worse, it was affecting American youth and families, and more funds were being allocated to do something about it. Basic data on the nature and extent of the problem were essential.

Because there were many unique aspects to the drug problem, the data that were needed could not simply be obtained by supplementing an existing survey. Major federal surveys at the time did not delve into sensitive personal issues, especially with youths. Asking about drug use meant asking about illegal behaviors that survey respondents would be reluctant to divulge, requiring special methodologies to obtain valid data. There were numerous drugs that were being used, the use was concentrated in young populations, and information on correlates such as attitudes were also important. Even if existing surveys could have been expanded to include new questions on drug use, the statisticians involved in those surveys opposed it because they were concerned about the effects on the quality of the data they were responsible for collecting. They knew that adding questions on a sensitive topic like illegal drug use could potentially affect interviewers, who would be uncomfortable administering the survey, and respondents, who might object to the topic and refuse to participate or alter the way they respond to other questions. To their credit, the statisticians were protecting the statistical integrity of those surveys. Years later, when NSDUH became well established, and much larger, we responded in the same way to requests from ONDCP, SAMHSA, CDC, and others asking us to add new topics to the survey.

The success of the two surveys conducted by RAC for the commission set the stage for the permanent institutionalization of the National Survey by the federal government. The commission surveys were well designed, using innovative techniques that were tailored to the content of the study, enhancing the willingness of respondents to agree to participate and to accurately report their illegal behaviors. One troubling aspect of the emphasis on encouraging respondent cooperation was the "cover-up" of the true goals and content of the survey until the interview was well underway. This practice was gradually removed from the survey design in subsequent years, in part because of ethical considerations and informed consent requirements. The design of the commission surveys also benefitted from the experience of a previously successful survey on a different sensitive topic, pornography. There was direct communication between the leader of the survey organization (Herb Abelson) and the lead scientist of the commission staff (Ralph Susman). Outside experts in substance abuse research were consulted. The surveys used state-of-the-art multistage sample designs, achieved a high response rate despite a very short data collection timeline, and applied appropriate weighting procedures to produce valid nationally representative estimates. Furthermore, the reporting of the results was straightforward, objective, and unbiased, despite the president's demand that the report support his point of view. The commission report included over 100 pages of details on the survey methodology, demonstrating transparency and full disclosure of the data limitations. The contribution and value of the two commission surveys to the field was evident when Louise Richards and her colleagues at NIMH and NIDA, including Director Robert DuPont, decided to continue to use the same methodology and conduct the surveys periodically.

However, the questionnaire itself and the way in which it was administered was modified in each survey. These changes were well-intentioned. The survey was in its infancy, and the research team discovered new problems and opportunities for improvement each time the survey was repeated, such as when they converted the alcohol module from interviewer to self-administered. They also revised the content each time the survey was conducted, presumably after assessing changing data needs or discovering new estimation methods such as the nominative technique. But as more years of data were added, and interest in tracking trends grew, the survey team faced a reality of the science of survey research: differences in survey procedures always carry a risk of affecting estimates and therefore comparability across time. Every change the survey team made to the design had the potential to cause a trend break, meaning an inability to use the survey data to describe whether drug use rates went up or down from one year to the next. To make matters worse, methodological research suggests that the impact of survey methods differences is generally greater for measurement of attitudes and sensitive behaviors, the two types of data that are a primary focus of NSDUH, than it is for other types of measures. And while it is perhaps widely known that changes in question wording or definitions affect survey responses, many researchers and other data users are surprised and sometimes disbelieving when they find that seemingly minor survey design differences can affect data comparability. Examples of survey design features that have been shown to affect estimates and response rates include:

- Mode of the interview, such as by telephone versus in person, or self-administered versus interviewer-administered, or paper versus computer.
- Setting, including classroom, home, or at a jail, and whether or not other persons are present during the interview.
- Characteristics of the interviewer, such as their demographics and interviewing experience.
- Question context, including the preceding questions, and the overall topic of the survey.
- Time of year data are collected.
- Sponsor of the survey and the organization collecting the data.
- Privacy protections promised.
- Incentive payments for participation.

These and other factors are important to consider at the survey design phase, and also when analyzing results. Some comparisons of estimates, particularly across years, should be made with caution or not at all where design changes have been made that could have affected those data. Also, survey design differences like those just mentioned must be considered when comparing results across different data sources.

With the results from several national drug surveys in hand by the end of the 1970s, government officials and researchers naturally turned to these data to assess trends in drug use. Despite the methodological changes in the survey, the data clearly showed substantial increases in illicit drug use during the decade, particularly marijuana use among young people. The ability to track changes over time has remained a priority for NSDUH data users to this day. Although changes to the survey have rendered some long-term comparisons invalid, with the application of crude adjustments to account for the methodological effects on the estimates, some long-term trend analyses are possible for certain specific measures and population groups. One such example is the trend in the rate of past thirty-day marijuana use among youths age 12–17, a widely used indicator of drug use in the United States. When President Nixon declared success in the drug war in 1973, the latest estimate for this indicator was 7 percent. By 1979 it had risen to 17 percent. Grouping the drug survey estimates by presidential administrations (see chart on the cover of this book) shows that the rate was highest during the Carter administration and lowest during the George H. W. Bush administration. Of course, attributing these shifts in rates of drug use to the changing policies of the administrations would not be appropriate. New policies or programs might take several years to be fully implemented and have any impact, and there are other factors that could influence trends in youth marijuana use, such cultural norms, attitudes, and perceptions of risk associated with marijuana.

3 Cocaine and New Directions for the Survey

President Reagan's Drug War

As the 1980s approached, there was growing public concern about the rapid rise in marijuana and other drug use among America's youth. Grass roots organizations, referred to as the parent's movement, led the resurgence of anti-drug sentiment.[1] These groups wanted the federal government to do more to prevent youths from using drugs, and they exerted considerable influence over government responses to the drug problem, especially after Ronald Reagan was elected president in 1980.[2] Although Reagan had campaigned on an anti-government ideology, with a goal of reducing the size of the federal government and its role in American citizens' lives, both he and First Lady Nancy Reagan embraced the parent's movement.

In his second press conference, President Reagan announced a plan for reducing non-defense personnel in the federal government, in part by shifting funding for some social programs to block grants. Funding would go directly to states, reducing the need for federal staff to manage these programs. During the press conference, Reagan was asked about his approach to the drug abuse problem.

Q. Mr. President, in light of what appears to be a growing concern about the drug abuse problem, especially among teenagers, what will your priorities be and specifically, do you expect to have a White House policy on drug abuse?

The President. Yes, I do. In fact, it can be stated as clearly as this: I think this is one of the gravest problems facing us internally in the United States. I've had people talk to me about increased efforts to head off the export into the United States of drugs from neighboring nations. With borders like ours, that, as the main method of halting the drug problem in America, is virtually impossible. It's like carrying water in a sieve.

> It is my belief, firm belief, that the answer to the drug problem comes through winning over the users to the point that we take the customers away from the drugs, not take the drugs, necessarily—try that, of course—you don't let up on that. But it's far more effective if you take the customers away than if you try to take the drugs away from those who want to be customers.
>
> We had a program in California—again, I call on that. We had an education program in the schools. We had former drug users who had straightened out. We found that they were most effective in talking to young people. You could go in, I could go in, anyone else and try to talk to these young people and tell them the harm in this and get nowhere. But when someone stood in front of them who said, "I've been there, and this is what it was like, and this is why I'm standing here telling you today," we found they listened.
>
> I envision whatever we can do at the national level to try and launch a campaign nationwide, because I think we're running the risk of losing a great part of a whole generation if we don't.[3]

A year and a half later, Reagan said more about his administration's drug abuse policies in a radio address from Camp David:

> In addition to the enforcement element, our strategy will also focus on international cooperation, education, and prevention-which Nancy's very interested in—detoxification and treatment and research. The mood toward drugs is changing in this country, and the momentum is with us. We're making no excuses for drugs—hard, soft, or otherwise. Drugs are bad, and we're going after them. As I've said before, we've taken down the surrender flag and run up the battle flag. And we're going to win the war on drugs.[4]

Two weeks after this radio address, Reagan officially announced his hard-line, zero-tolerance approach in the war on drugs. There would be substantial increases in funding for interdiction, law enforcement, and prisons, and harsh penalties for people using illegal drugs. The US military would be authorized to assist in drug law enforcement. The emphasis shift was evident in the annual budgets. The proportion of the total federal drug control budget that was allocated to supply reduction increased from 56 percent in 1981 to 71 percent in 1987.[5]

The most prominent demand-related activity of the Reagan Administration was Nancy Reagan's "Just Say No" campaign. NIDA had developed a Drug Abuse Prevention Media Campaign, which targeted children age 12–14, and promoted abstinence, parental involvement, and resistance to peer pressure. Mrs. Reagan was visiting Longfellow Elementary School in Oakland, California, where she watched a video from the media campaign in which a boy was asked what he would do if asked to take drugs. The boy replied "I would say no." When asked later by one of the students what to do if offered drugs, Mrs. Reagan responded "Just say no."[6] The phrase caught on, and Just Say No clubs within schools and school-based drug prevention programs became common nationwide.

Funding for the federally supported drug treatment system was reduced and restructured early during the Reagan Administration. Responsibility for management of the treatment system was shifted from NIDA to the states, with funding provided to states through a block grant. One consequence of this change was the loss of an important research data set, the Client Oriented Data Acquisition Process (CODAP). From 1973 until 1981, the programs that received federal funds were required to report data to NIDA on each admission and discharge. The data included dates of admission and discharge, demographic characteristics, drugs used by patients, outcome of treatment, and locations of the clinics. NIDA analysts compiled these data and used them to monitor the use of the funds, and track trends in drug treatment. With the block grant, there was no longer any requirement for programs to submit the data. CODAP collapsed. National client level data was not available again until after the Anti-Drug Abuse Act of 1988 specified new data reporting requirements, resulting in the Treatment Episode Data Set (TEDS).

In this era of changing drug abuse attitudes, policies, and data, the management of the National Survey on Drug Abuse was also in transition. The survey was conducted only three times during the eight years of the Reagan Administration, and each survey would have a different government project officer and contractor. There would also be changes in the survey design and reporting of results. Change can be difficult and challenging, and everything did not proceed smoothly.

1982 National Survey on Drug Abuse

NIDA planned for the next survey to be carried out in 1981. Louise Richards and Joan Rittenhouse had successfully shepherded the project through four rounds of data collection and analysis. They had improved the questionnaire and maintained most of the key measures, resulting in a data series that could be used to analyze changes in drug use patterns during a time of unprecedented increases. But in 1980, NIDA shifted responsibility for the National Survey to its Division of Medical and Professional Affairs (DMPA). Dr. William Spillane became the new project officer. Under the purview of this division, the focus of the survey shifted towards psychotherapeutic drugs – sedatives, tranquilizers, amphetamines, and analgesics. The goal was to add new questions on medical use of these prescription drugs and on consequences of medical and nonmedical use of these drugs. Consequences included tolerance, withdrawal, difficulty cutting down and need for professional help to cut down, family, school and work problems, accidents, and overdoses.

Spillane solicited ideas from Richards, who cordially provided thoughtful feedback. Louise suggested that the sampling frame be expanded beyond households, that data on quantity of drugs used (e.g., number of joints, pills, injections) should be collected, and that proposed medical use questions be cut back. She also suggested that Spillane continue to collaborate with her branch, and that he encourage the contractor to apply to her branch for a grant to do further methodological development work.[7] Louise wanted to keep the survey in her life. Later, she reviewed a draft of the questionnaire and saw that DMPA planned to drop all cigarette questions, to make room for the new psychotherapeutics questions. She sent a memo to Spillane strongly urging him to retain the cigarette module.[8]

Early in 1981, a draft questionnaire was sent to NIMH for review. NIMH strongly objected to the proposed questions on medical use and consequences, saying the approach was flawed and imprecise. They wanted NIDA to limit the survey to the collection of illicit drug data and *nonmedical* use of psychotherapeutics, and keep the questionnaire the same as in the 1979 survey, to detect trends. Their response stated "The key to the success of the survey as a barometer of changes is the principle of replication." NIMH also said that NIDA did not have the expertise to delve into medical issues: "The whole subject of the medical use of psychotherapeutic drugs is and has been the proper province of NIMH," and these kinds of assessments should be done by those "who are most centrally involved with treatment research and who are most experienced in the development and conduct of general population surveys on the subject of medical use: namely, NIMH." NIDA disagreed with many of the comments from NIMH, viewing them as sarcastic, arrogant, and self-serving.[9] Soon after the NIMH review, NIDA convened an external technical review committee to assess the proposed questionnaire. The panel of experts generally endorsed the new questions, but recommended some modifications, which NIDA made.[10]

Sampling and household screening procedures for 1982 were the same as in 1979. But there were numerous changes to the questionnaire. The questions on medical use of psychotherapeutics were included, in the interviewer-administered part of the survey. A new self-administered module on consequences of medical and nonmedical use of these drugs was also added. The nonmedical pill use modules were modified to obtain information on circumstances of use and indicators of dependence, and changed from interviewer-administered to self-administered, as four separate modules (sedatives, tranquilizers, stimulants, and analgesics) placed between the marijuana/hashish and cocaine modules. As suggested by Louise Richards, the cigarette questions were retained, in the interviewer-administered portion of the interview. But the inhalants module was not included in the survey. The alcohol, marijuana, cocaine, hallucinogens, and heroin answer sheets were all different in 1982 than in 1979. Different items were included. For some of the items that were in both years, the ordering and specific wording was changed in 1982. And a key item that was included with the same wording in 1979 and 1982, the question on recency of use, had the same wording in both years, but the response categories were changed. The seven categories shown in the 1979 version were collapsed into four categories in 1982. Notably, the "past week" and "past month" options were collapsed into "past month (30 days)." All of these changes had the potential to affect comparability of the 1979 and 1982 estimates.[11] A contract was awarded to GWU and RAC again, to conduct the 1981 NSDA. But the contract started on September 25, 1981, too late to conduct the full survey in 1981. The survey would have to be pushed back to 1982. However, an experiment testing the impact of incentives (cash gifts to respondents) would be done in November and December of 1981, and the 313 interviews from this study were combined with the 5,311 main survey interviews completed from March through July 1982, for a total sample size of 5,624. The interview response rate was 84 percent.

The 1982 results reported a reversal of the upward trends in drug use seen in prior surveys. There was a decline since 1979 in the percentage of young adults using marijuana, cocaine, hallucinogens, and alcohol in the past month. Marijuana and alcohol use declined among youths. Although estimates for nonmedical prescription drug use were higher in 1982 than in 1979, the report was vague about whether this was a real finding. Acknowledging the change from interviewer- to self-administered in questions on nonmedical psychotherapeutic use, the Main Findings report says "The apparent 1979 to 1982 increase in nonmedical use of these drugs in the youth population may be due either to a real increase in use or to a greater tendency to report use on the 1982 self-administered form." But there was no acknowledgement of the potential effects of the changes in the answer sheets for all of the other drugs. The report stated "Since data from earlier surveys in this series are readily available and since the form and method of questioning have been reasonably comparable over the years, it is possible to provide information on trends in illicit drug use over more than a decade."[12] As discussed in Chapter 2, the large sampling errors in these early surveys and a reliance on statistical significance testing offered some protection against erroneous trend interpretations due to survey design changes. Concerns about trend breaks due to minor questionnaire changes increased in later years (1991 and later) in which the sample size was much larger.

Consolidation of National Drug Abuse Data Systems

Dr. William Pollin was NIDA's Director of Research under Dr. DuPont, then became NIDA Director after DuPont's departure in 1979. Like his predecessor, Pollin recognized the value of having a variety of statistical data sources for research and policy development. But in 1980, NIDA's principal data sources were managed by different offices. The NSDA was housed in NIDA's Division of Medical and Professional Affairs. The Monitoring the Future study (MTF), an annual survey of high school seniors, was conducted under a grant through the Division of Research. The Drug Abuse Warning Network (DAWN), which captured data on drug abuse related incidents reported by hospital emergency rooms and medical examiners, was managed by the Drug Enforcement Administration (DEA) of the Department of Justice. NIDA's Forecasting Branch collaborated with DEA to obtain DAWN data, and produced reports based on their analysis of these data. Treatment-related data were managed by NIDA's Division of Data and Information Development (DDID). The treatment data sets included the National Drug and Alcoholism Treatment Unit Survey (NDATUS) and CODAP. NDATUS was a survey of all known substance abuse treatment facilities, including private and publicly funded organizations. Edgar Adams, the Associate Director for Epidemiology in DDID, recognized the value in having a single division oversee all of NIDA's data systems. He envisioned this centralization bringing a new synergy to the drug epidemiology activities in NIDA. It would facilitate collaboration among a core group of statisticians, epidemiologists, and other researchers in NIDA, and lead to stronger analyses that combine data from multiple sources. Adams presented his epidemiology plan to NIDA Director Pollin. Discussing the plan in a meeting with National Survey pioneer Herb Abelson, Pollin said that ten years ago, there wasn't anything like this collection of data sources for making decisions, and that if there had been, it is likely that some of the decisions made would have been different and would have been dealt with more carefully and thoughtfully.[13]

Pollin approved the plan. The NSDA and MTF were moved into the DDID, where CODAP and NDATUS already resided. The division was renamed Division of Epidemiology and Statistical Analysis (DESA). An interagency agreement between DEA and NIDA facilitated the transfer of DAWN to DESA. For the first time, the National Survey was managed by what one might call a "data group" – an organization with a focus and primary responsibility for managing a large program of statistical data collection and analysis.

1985 National Household Survey on Drug Abuse (NHSDA)

The next survey, to be conducted in 1984 and renamed the National Household Survey on Drug Abuse, was the responsibility of DESA. During the planning of the survey in 1983, epidemiologist Dr. Beatrice Rouse was hired to take over as project officer. She would work under the supervision of Louise Richards, who had been reassigned to DESA to serve as Chief of the Epidemiologic Research Branch. Although in a different branch, I was assigned to be alternate project officer.[14]

Prior to the 1984 survey, NIDA carried out an extensive review of the survey design and questionnaire. Beginning in 1982, consultants were identified and asked for recommendations regarding all aspects of the survey design and questionnaire. Consultants included NSDA contractors Herb Abelson and Ira Cisin, along with other distinguished statisticians (Leslie Kish and Joe Waksberg) and substance abuse researchers (Lloyd Johnston, Richard Clayton, Denise Kandel, and Robin Room). In June 1983, NIDA Director Pollin decided that more time was needed to review the survey design. He postponed the next survey until 1985. Final recommendations, submitted to NIDA early in 1984, were critical of the number of changes that had been made to the questionnaire, methods, and time of year of data collection in prior years, causing disruptions in the survey's capability to track trends. The suggestion for 1985 was to "try to get it right and stick with it." The consultants recommended that a set of core questions on drug use be placed in the beginning of the questionnaire, and any new questions added in future years should be placed at the end of the interview, so there would be no changes in the context under which the core questions are presented to respondents. They concluded that the survey should be conducted every two or three years, during the same months of the year each time. The group studied the questionnaire in detail, and made numerous specific suggestions, most of which were implemented in the 1985 survey.

The contract for the 1985 survey was awarded in September 1984. For the first time in the survey's history, George Washington University and Response Analysis Corporation would not be involved. The winning bid for this round was from the Institute for Survey Research (ISR) at Temple University, led by principal investigator Dr. Leonard LoSciuto. Analysis and report writing would be done by a subcontractor, the University of Kentucky. Co-principal investigators at Kentucky were Dr. Richard Clayton and Dr. Harwin Voss.

From start to finish, there was conflict and animosity between the contractor and the DESA officials. At the first contract meeting, on October 12, there was already a disagreement. The contract called for a spring 1985 survey, to match the 1982 spring data collection period. However, ISR requested that the survey be carried out in the fall. Based on their experience, and reviewing the upfront work that needed to be done before data collection could begin, ISR reasoned that there was not enough time to put the survey into the field in the spring. But NIDA would not change the schedule, saying there was a great need to have the data as soon as possible. To meet this tight schedule, there were requests to the Office of Management and Budget (OMB) over the next few months for expeditious review of the clearance package; a compressed schedule of questionnaire testing; a request to the Census Bureau for rapid access to maps needed for sampling; and exploration of mechanisms for quick turnaround of questionnaire printing jobs. There were also conference calls with trusted NIDA consultants and leading drug epidemiologists Ira Cisin, Lloyd Johnston, and Denise Kandel to get their recommendations regarding the schedule. And despite the tight schedule, as late as January 18, NIDA Director Pollin requested that a major new component be added to the project: a follow-up study, in which respondents would be interviewed a second time at a later date. This idea was deemed not feasible, as were several requests from states that wanted to "piggyback" on the survey, adding funds to the contract to expand the sample sizes in their states. It was barely possible to complete all preparations for a spring survey without this added sampling and field work. Eventually, the first completed interview was achieved in June 1985.

The 1985 questionnaire had some differences from previous surveys, but most of the drug use questions were comparable to the ones employed in 1982. As suggested by the consultant panel, the inhalants section was returned to the questionnaire, and questions on friends' heroin use, for estimating prevalence using the nominative technique (see Chapter 2), were dropped. New questions ascertaining the lifetime occurrence of dependence and abuse symptoms for each substance were

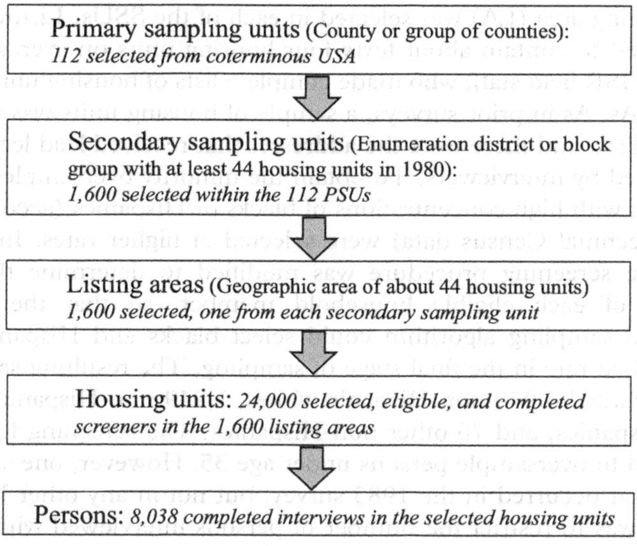

Primary sampling units (County or group of counties):
112 selected from coterminous USA
Secondary sampling units (Enumeration district or block group with at least 44 housing units in 1980):
1,600 selected within the 112 PSUs
Listing areas (Geographic area of about 44 housing units)
1,600 selected, one from each secondary sampling unit
Housing units: 24,000 selected, eligible, and completed screeners in the 1,600 listing areas
Persons: 8,038 completed interviews in the selected housing units

Figure 3.1 Multistage sample for 1985 survey.

added, and a new module on treatment for substance abuse problems was inserted. Also, due to the co-sponsorship of NIDA's sister agency NIAAA, new questions on problems caused by alcohol use were added. Although the consultants had recommended keeping the questions on medical use of drugs, these items were not included in the 1985 survey. The number of questions on nonmedical use of psychotherapeutics was reduced, a change the panel had recommended, although questions on specific problems resulting from use of these drugs were retained. Finally, there was a new module of questions on perceived risk of harm associated with various drug-using behaviors.

As in prior surveys, a multistage area probability sample was selected. But NIDA specified a new requirement to oversample black and Hispanic populations. This new sampling approach would make it possible to show more detailed race/ethnicity categories in the survey's reports. Categories shown in prior reports ("White" and "Black and other races") were not useful. The oversampling of blacks and Hispanics became part of the NHSDA design for the next thirteen years. Of course, the RAC sample of interviewing locations used in each prior survey was no longer used. Instead, it was ISR's National Sampling Frame of 112 primary sampling units (PSUs) and 5,200 secondary sampling units (SSUs) from which the 1985 sample of 1,600 SSUs was drawn (see Figure 3.1).

One listing area (LA) was selected in each of the SSUs. Listing areas, constructed to contain about forty-four housing units on average, were visited by ISR field staff, who made complete lists of housing units in the sample LAs. As in prior surveys, a sample of housing units was selected from the lists, and these were the addresses that received lead letters and were visited by interviewers. To obtain the minority oversamples, PSUs and SSUs with high concentrations of blacks or Hispanics (according to 1980 Decennial Census data) were selected at higher rates. In households, the screening procedure was modified to determine the race/ethnicity of each eligible household member, so that the within-household sampling algorithm could select blacks and Hispanics at a pre-specified rate in the final stage of sampling. The resulting sample of 8,038 included 3,979 non-Hispanic whites, 1,949 non-Hispanic blacks, 2,034 Hispanics, and 76 other non-Hispanics. The screening form also continued to oversample persons under age 35. However, one sampling change that occurred in the 1985 survey, but not in any other National Surveys, was to restrict the number of persons interviewed within each household to no more than one. Despite NIDA's requirement for spring data collection, interviewing did not begin until June. It was completed in December. The interview response rate was 83 percent.

Because of NIDA's desire to have the survey completed as soon as possible, much of the interviewing took place in the summer. Data collection is typically more costly in the summer than in fall or spring, because people are less likely to be at home, requiring more return visits. Costs exceeded the original budget, and in early 1986, right after the survey was completed, principal investigator LoSciuto informed project officer Rouse that his projections of costs for all of the remaining work, which involved data editing, creating data files, and writing reports, showed that the budgeted amount would not be sufficient. At this point, NIDA's contracts office decided, after discussion with Rouse and LoSciuto, that to complete the project, pay for the cost overrun, and prevent any more cost increases, the remainder of the contract would be converted to a fixed-price contract, for just over a million dollars. With the data collection phase over, it was assumed that costs would be more predictable.

But once the data processing and analysis began, there were more problems. Preliminary analysis showed the estimated number of lifetime cocaine users was significantly lower than the number in 1982. This indicated a problem. The number should have been higher in 1985 than in 1982, because there were many new users in the population during the three years between the surveys, and all lifetime users in 1982 would still be lifetime users in 1985. The contractor's investigation of the anomaly determined that further editing of the 1985 data was needed. After the edits were done, revised estimates were more realistic. Then in late summer 1986, NIDA Director Pollin asked DESA acting director Edgar Adams to review a draft report that included some 1985 NHSDA estimates. Edgar was surprised to find that the contractor had provided the data to Pollin without telling the project officer (Rouse) or anyone else in DESA. This was a violation of the contract, which stipulated that all contract work and data releases must be approved by the project officer.[15] In addition, the estimates given to the NIDA director were incorrect. Despite these cost overruns, errors and contract violation, NIDA achieved its goal of getting the results out quickly. The NHSDA results, contained in a few pages of highlights and a set of data tables, were released by HHS Secretary Otis Bowen, MD, at a press conference on October 9, 1986. In his remarks, Bowen described mixed results, with cocaine use increasing and marijuana use decreasing between 1982 and 1985. However, he mainly focused on a separate telephone survey done by NIDA (National Survey of Beliefs and Knowledge about Illegal Drug Use) that showed that 73 percent of Americans describe illegal drug use as one of the most serious problems facing the country. Reviving the war rhetoric, Bowen said:

> Victory over illegal drug use will not come automatically or easily. We must not over-promise. But I believe these surveys are an indication that victory is possible. Just as the Normandy invasion in World War II could only be launched when there were enough men, enough materiel, enough information and enough success on other fronts, American attitudes toward drugs have now reached a critical point where a full-scale assault is possible.

Donald Ian MacDonald, ADAMHA administrator, also spoke at the press conference, describing the results of the NHSDA in more depth. He highlighted the increase in the number of past month cocaine users, from 4.2 million in 1982 to 5.8 million in 1985, as a great concern. Although not emphasized at the press conference, a new measure of illicit drug use appeared in the press release and in the brief highlights. Estimates of "any illicit drug use" were constructed, defined as use of marijuana, cocaine, heroin, hallucinogens, inhalants, or nonmedical use of prescription drugs. This new indicator became a standard measure of the prevalence of drug use in the population for all subsequent National Surveys. The concept was not new. "Any illicit" estimates had been presented in MTF reports, but with a slightly different definition that did not include inhalants as an illicit drug.[16] The NHSDA press release reported that 23 million people, or 12 percent of the population age 12 and older, were currently using illicit drugs (used in the past thirty days).

The Population Estimates report from the 1985 survey was published in 1987, but the Main Findings report was lagging.[17] NIDA's goal was to have consistency between these two reports, but drafts of the Main Findings were showing estimates that were different from those already published in the earlier report. The contractor was asked to explain. LoSciuto replied that the differences were mainly due to inconsistent variable definitions used by ISR (for Population Estimates) and Kentucky (for Main Findings), associated with how missing data were handled. There were communication problems between the contractor and subcontractor. But other concerns emerged with the draft Main Findings report, and other contract deliverables were late. In the midst of these challenges, another contract violation was discovered. Voss and Clayton had done an unauthorized analysis of the 1985 data and submitted it to a peer-review journal in December 1986. Neither the government project officer Rouse nor the principal investigator LoSciuto knew about it until after the article appeared in the journal in 1987. And once again, the analysis and estimates in the paper were incorrect. The analysis used unweighted data, ignoring the oversampling of minorities and young people built into the sampling plan, resulting in biased estimates that were not nationally representative. Furthermore, the paper did not provide sampling errors or tests for statistical significance, in accordance with standard statistical practice.[18]

The working relationship between Dr. Rouse and the contractor had deteriorated. There were long delays, poorly communicated instructions from the government, unacceptable draft reports, contract violations, and scolding letters. Early in 1988, I was asked to help resolve the situation. My task was to review the Main Findings report and try to get it finished. Jeanne Moorman, a statistician I had hired to work on the survey, also helped. Our review revealed some troubling statistical issues. First, there was no testing for statistical significance to support the narrative describing demographic differences in drug use rates. So for example, although the rates for lifetime marijuana use for the sexes appeared to be different (25.4 percent for males age 12–17 and 21.6 percent for females age 12–17), the sampling errors for these estimates were so large that the difference was not statistically significant. This meant that there was not enough evidence to conclude that teenage boys were more likely than teenage girls to have tried marijuana. To be fair, significance testing in this manner had never been done in any prior Main Findings reports. But it was time for that to change. A related concern about the report was how the sampling errors were calculated. In the draft report, sampling errors were computed based on the assumption that the design effect (the impact of the complex sampling design on the precision of estimates) was the same for every estimate in the report.[19] This assumption greatly simplifies the calculation of sampling errors and the testing for statistical significance, but for the NHSDA it was a questionable assumption. This simplified approach had been used in previous rounds of the survey, and may have led to inaccuracies because it did not account for the oversampling of young people, which can increase the design effect for estimates for combined age groups. Now that the 1985 survey also oversampled blacks and Hispanics, this shortcut in estimating sampling errors was a bigger concern. So we suggested that separate design effects for different population subgroups be applied, to improve the accuracy of the computed sampling errors. Additionally, the review revealed an unusual recoding of respondents' ages. The contractor had coded respondents' ages as of July 1, 1985 – the same day for every respondent – instead of their age on the day they were interviewed. This was a change from previous surveys, and in our view, incorrect. Recoding to the July 1 date was done using the date of birth reported by each respondent, and was apparently done so that the sample would coincide with the population counts used in estimating the number of drug users. These counts, obtained from the Census Bureau, reflect the US population on July 1, 1985. But the household eligibility determination and sampling were based on the age on the date of the screening interview visit. More importantly, the data on drug use and other characteristics all reflect the respondent on the day of the interview. A shift in respondents' ages could affect statistical relationships and distort patterns of use by age.

After Beatrice told the contractor to make these corrections, they immediately sent a letter to the NIDA Director Schuster, complaining about NIDA's management of the survey.[20] They said that the age recoding would require a great deal of work, including recalculating the analysis weights, and would result in estimates that differ from those already released, and that they did not have the money or the time to do this work. Of course by this time, the remaining work was to be done under a fixed-price contract.

Eventually the contractor did the work, and delivered an improved Main Findings report, published in 1988. A final data tape, documentation, and other deliverables were completed, as the contract lingered into 1989, and finally came to a welcome end. By this time, the 1988 survey had already been conducted, through a separate contract, with a new contractor and project officer.

1988 National Household Survey on Drug Abuse

In May 1987, the NIDA contracts office issued a request for proposals to conduct the 1988 and 1990 NSDHAs. The contract stipulated data collection in the spring for both surveys. As in 1985, NIAAA co-sponsored these two surveys, contributing about 20 percent of the costs. The Department of Education also gave NIDA funds to support the surveys.

Three strong proposals were submitted and were reviewed twice by an independent, external review panel. Due to budget problems in NIDA, it was decided to delay the contract award until the next fiscal year (after October 1, 1987). I had no involvement in any of this pre-contract activity for the 1988 survey. But not long after the second review panel meeting in January 1988, I was informed that I would be the project officer for the contract. This was a surprise to me, especially since in addition to taking over responsibility for the project, I would be Chief of the Survey Research and Reports Section. I had mixed feelings. I had never aspired to become a supervisor, and was a bit apprehensive about it. But I felt prepared to oversee the contract. I had worked on all aspects of large surveys in my previous job at NCHS, and was familiar with the NHSDA from analysis I had done using the data files and the review of Main Findings. I had also been project officer on a small validation study that duplicated the NHSDA's methods.[21]

I was handed the three proposals, the external panel reviews, bidders' responses to questions, and other relevant paperwork on the procurement activity up to that point. I had not attended either of the review meetings, nor had I seen any of the proposals or documentation before. My first task was to review all of the materials and help the contracts office with final negotiations with the three bidders. As project officer I had to make a recommendation, with clear rationale based on the materials I was given, on which firm should win the contract. Any of the three could have successfully carried out the two surveys. Scores from the review panel ranked them essentially equal, and cost proposals were also similar. Two of the bidders had already conducted the NHSDA. But I recommended the third: Research Triangle Institute (RTI). It was probably my grounding in large-scale survey research that swayed my thinking. Of the three bidders, RTI seemed to have the most depth in terms of sampling, estimation, methodology, and statistical analysis. They were a big company with a cadre of survey statisticians that could potentially be called upon throughout the contract, as needed. The contracts management office agreed, and awarded the contract to RTI on April 5, 1988. The RTI team was led by Project Director Thomas Virag. Once again, the original schedule had to be revised due to the delays in the award. The 1988 NHSDA data collection would take place in the fall, as would the 1990 survey.

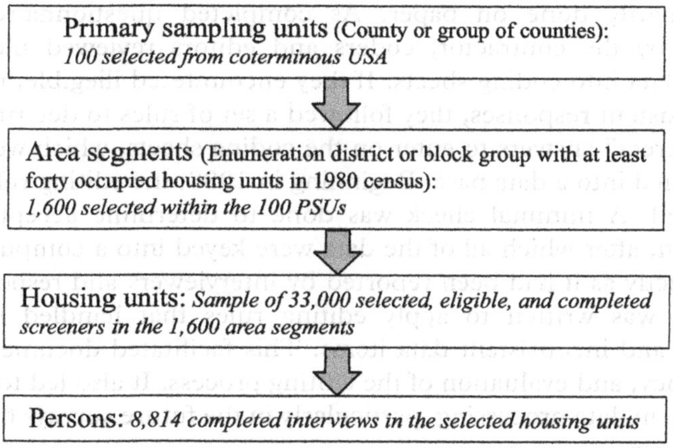
Primary sampling units (County or group of counties):
100 selected from coterminous USA
Area segments (Enumeration district or block group with at least forty occupied housing units in 1980 census):
1,600 selected within the 100 PSUs
Housing units: Sample of 33,000 selected, eligible, and completed screeners in the 1,600 area segments
Persons: 8,814 completed interviews in the selected housing units

Figure 3.2 Multistage sample for 1988 survey.

The design and data collection procedures of the 1988 survey were similar to the 1985 design and procedures. The sample design is shown in Figure 3.2.

It was a multistage area probability sample, with oversampling of blacks, Hispanics, and young people. With a new contractor, it used an entirely new sample of primary sampling units (100) and area segments (1,600) within those PSUs. The limit of only one sample person per household that had been in place for 1985 was dropped. Either zero, one, or two respondents could be selected in each sample household in 1988. Data collection began in September and was expected to be completed by the end of the year, but as the end of the data collection approached it became clear that because of difficulties in obtaining the required response rate, RTI would fall far short of the required sample size of 9,400. Additional sample addresses were selected and the field period was extended until February 28. The final sample size was 8,814. The interview response rate was 74 percent.

The 1988 questionnaire was nearly identical to the 1985 questionnaire. To address emerging data needs, new questions on crack use were inserted into the cocaine module, and questions about needle sharing were also added. The module on drug abuse treatment was dropped. Also, questions on lifetime symptoms of drug abuse and dependence were changed to capture information about the past year.

Several improvements related to editing, imputation, and the estimation of standard errors suggested by RTI were incorporated into the NHSDA for 1988. Prior to 1988, editing of drug use and other variables was primarily done on paper. As completed questionnaires were received by the contractor, coders and editors reviewed them and entered data into coding sheets. If they encountered illegible, missing, or inconsistent responses, they followed a set of rules to determine the final "correct" answers to enter on the coding sheets, which were then keypunched into a data base. Beginning in 1988, the editing rules were automated. A minimal check was done to determine acceptance of each form, after which all of the data were keyed into a computer data base, exactly as it had been reported by interviewers and respondents. Software was written to apply editing rules that handled missing, illegible, and inconsistent data items. This facilitated documentation, consistency, and evaluation of the editing process. It also led to greater efficiency in data processing, particularly in the future surveys that were much larger.

While editing improved the quality of data by ensuring consistency and acceptable responses, it did not eliminate all missing data. To further address missing data after editing, statistical imputation became an integral component of the data processing and variable creation starting in 1988.[22] Imputation uses a statistical model to make a reasonable guess or prediction of what the true response for a missing item should be. That predicted value is inserted in place of the missing value. Imputation is commonly used in large-scale surveys, and typically results in more accurate and less biased estimates than simply ignoring the missing data or excluding cases with missing data from estimation. For the 1988 NHSDA, a hot-deck imputation method was applied to a handful of demographic variables and most of the major drug use indicators. Hot-deck imputation of a key variable involves sorting records for all respondents according to a set of other variables for which there is no missing data, and which are correlated with the key variable. This sorting creates groups of respondents that are "similar" in terms of these variables. A respondent with missing data for the key variable is then assigned an imputed value by taking a "donated" value for the key variable from the next respondent (a donor) in the sorted list that has non-missing data for the key variable.[23]

A third analytic enhancement in 1988 involved the calculation of estimated standard errors for the survey estimates of drug use and other variables. Prior National Survey reports employed a short-cut, simplified method that assumed that a single basic formula could be used to approximate the standard error for any estimate. However, because use patterns can vary a great deal across different drug types, this assumption may not be correct. Accurate estimates of standard errors are essential for correctly interpreting the results of the survey. Standard errors guide analysts in determining if estimates have adequate precision and if the prevalence of use of a drug is increasing, decreasing, or remaining level, and whether rates differ across population subgroups. Beginning in 1988, NHSDA tables and reports computed estimates of standard errors separately for every individual prevalence estimate.

Cocaine, Crack, and the Drug Czar

The 1982 National Survey showed that the steep increases in marijuana use seen during the 1970s had ended and use among young people was declining. With the two or three year intervals between surveys, it is difficult to pinpoint the peak year for youth marijuana use, but it was probably 1978 or 1979.[24] The decline continued throughout the 1980s, as the 1985 and 1988 NHSDAs showed. Although the survey showed that cocaine use among youths and young adults had leveled off in 1982 and 1985, cocaine emerged as a major drug of concern. Use was increasing among older adults. From 1979 to 1982, the rate of past year cocaine use nearly doubled among adults age 26 and older, partly due to the aging of the cohort of users that was 23–25 years old in 1979 and 26–28 years old in 1982. By 1985, one-quarter of the US population age 18–34 had tried cocaine in their lifetime. Emergency department visits involving cocaine abuse increased more than seven-fold between 1980 and 1984. Cocaine related deaths increased. The CODAP data were severely degraded during this period, but what remained of the data set indicated increases in admissions to treatment for cocaine problems. With the consolidation of data sets in one NIDA division, analysts were able to produce a series of research papers on cocaine use in the 1980s. These papers described the patterns and trends in cocaine use and consequences, providing important baseline information for policy-makers formulating responses to the problem.[25]

Often when analysts compile all available data to study a particular problem, they discover data gaps, or specific information that is needed for the study, but does not exist. The demise of CODAP created such a data gap. Attempts to use National Survey data to track the emerging crack problem exposed another gap. The survey was not done frequently enough and it was unable to address emerging drug abuse behaviors and patterns. When NIDA began to realize and understand the significance of crack, a cheaper, smoked form of cocaine, the 1985 questionnaire had already been finalized. The next survey would not be conducted until 1988. Thus, the first data on crack use in the general population would not be available until 1989. The 1985 questionnaire did ask about route of administration of cocaine, so it was possible to estimate smoked cocaine. But the questionnaire did not distinguish crack use from free-basing, another way of smoking cocaine.

Despite the statistics, the hazards of cocaine use may not have been widely understood by the American public, just as had been the case nearly 100 years before. Media portrayals of the cocaine problem were mainly about the violence associated with the drug trade, and the devastating effects of crack in inner cities. The 1985 NHSDA showed that only about one-third of persons aged 12–34 thought that use of cocaine one or two times posed a great risk of harm. But when it comes to changing minds, statistics and public service announcements are no match for real-life tragedies that happen to celebrities and heroes. The news on June 19, 1986 that basketball star Len Bias had suddenly died, overdosing on powder cocaine – not crack – stunned the nation. Bias, 22, had just completed a stellar career at the University of Maryland. A 6 foot 8 inch power forward, the All-American was exceptionally strong, but had a soft touch as an outside shooter. Just two days before his death, he had been selected by the Boston Celtics as the overall number two pick in the NBA draft. The initial media reports, later determined to be inaccurate, said that Bias had not been a regular user of cocaine, and that he had just begun experimenting with the drug. The impact of the Bias overdose was significant for a number of reasons. First, it was shocking because of the sheer strength and physical invulnerability of Len Bias. Could cocaine be so dangerous and unpredictable that it could kill such a vigorous, healthy, strong athlete? Secondly, the overdose occurred right outside Washington, DC, only a few miles from the Capitol. The president and Congress noticed, especially the Speaker of the House, Tip O'Neill, who represented Boston. Attention to the cocaine problem heightened eight days later when football player Don Rogers of the Cleveland Browns died of a heart attack caused by his cocaine abuse, the day before he was to be married.

Most politicians in the 1980s agreed the war on drugs was necessary. Regardless of their political party, they did not want to appear to be soft on drugs. In 1986, Democrats, led by Speaker O'Neill, saw a political opportunity to show they could be tough on drugs. To address the cocaine problem, and specifically in response to the overdose deaths in June, Congress passed two new laws associated with the drug war. They quickly passed the Anti-Drug Abuse Act of 1986. This bill was introduced on September 8, passed by both the House and Senate, and signed into law by President Reagan on October 27, 1986. The law increased penalties for drug distribution and possession, and required mandatory minimum sentencing for drug offenders.

Two years later, the Anti-Drug Abuse Act of 1988 was passed by both houses of Congress, and signed into law by President Reagan on November 18, 1988. This law had a huge impact on the NHSDA and other data systems. The law called for new data collection and increased research funding. It also established the White House Office of National Drug Control Policy (ONDCP). The director of ONDCP would be the drug czar. Soon ONDCP would begin to directly influence the Household Survey.

Release of the 1988 NHSDA Results

Other than a major cost overrun and the need to extend the data collection period into 1989, the 1988 survey had progressed smoothly and the data were ready for release a few months after data collection ended. On July 31, 1989, HHS hosted a press conference to announce the results. Louis Sullivan, MD, the new HHS Secretary under President Bush, and NIDA director Dr. Charles Schuster spoke. The other key speaker was William Bennett, the new drug czar.

Sullivan reported that overall drug use was down significantly. The number of past year illicit drug users ("any illicit use") declined from 37 million in 1985 to 28 million in 1988. Even cocaine use had declined, from 12.2 million to 8.2 million past year users. The trends were encouraging, but the numbers of users were still too high, Sullivan stated.

The decrease in cocaine use may have been partly the result of the public's realization of how dangerous cocaine use can be, due to the highly publicized overdose deaths of Len Bias and Don Rogers in1986. Figure 3.3 shows that the percent of the population age 12 to 34 believing there was great risk of harm in using cocaine once or twice increased by about twenty percentage points between 1985 and 1988.

The Monitoring the Future study shows how immediate and lasting the impact on youth was. The percentages of 12th graders (interviewed each spring) believing there was great risk of harm in using cocaine were 35.7 in 1984, 34.0 in 1985, and 33.5 in 1986. Then the rate jumped to 47.9 in 1987, and continued to rise after that (51.2 in 1988, 54.9 in 1989, and 59.4 in 1990).[26]

Typical for these press conferences, there was a mix of good and bad news. The political leaders like to take these opportunities to both cite their success and remind everyone that more needs to be done (i.e., continued funding support). We're winning the war, but we need more troops. The wide range of estimates the survey produces each year always provides data for both claims. The NHSDA estimated that the number of people who used cocaine on at least a weekly basis during the past

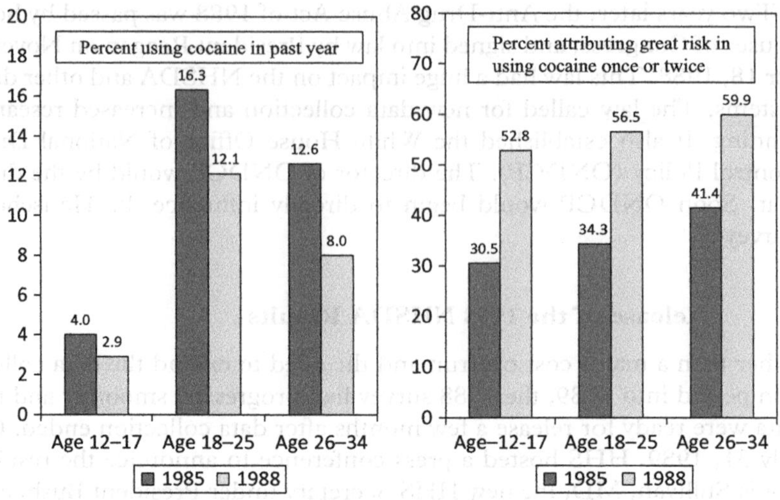

Figure 3.3 Cocaine use and perceived risk by age, 1985 and 1988.

twelve months was 647,000 in 1985 and 862,000 in 1988. Because these estimates are projected from small samples and represent rare behaviors (less than a half percent of the population used cocaine weekly), they are unreliable. In other words, sampling errors for these estimates are relatively large. Consequently, the difference between them is not statistically significant. But the new drug czar wanted to make a point that there had been no progress in reducing heavy cocaine use, calling this "a shocking and unacceptable jump in just three years." He tied the "increase" to the appearance of crack in inner cities, and stated "The fact that first sightings of crack are now reported almost daily in suburbs and rural areas around the country is an alarming portent for the future."

This first experience with ONDCP participation in the release of the National Survey results and their willingness to ignore statistical principles was an alarming portent for the future of our division, which was reorganized and renamed the Division of Epidemiology and Prevention Research (DEPR) in 1989. Listening to Bennett's speech that day, I wondered if we would be facing other conflicts with ONDCP over statistical issues. There are always concerns about senior officials interpreting data to promote particular policies or political agendas, but the tone of Bennett's speech was more troubling than usual. Soon we would find that this concern was warranted, but not just because of how the ONDCP might use the data. Soon ONDCP would be telling us to change the design of the survey.

Notes

1 The National Families in Action, the Parents' Resource Institute on Drug Education, the National Family Partnership, and the Partnership for a Drug Free America helped form thousands of individual community parent groups across the nation.

2 The influence of the parent's movement remained strong throughout Reagan's presidency. Donald Ian MacDonald, MD, a national leader in the parent's movement, was appointed by Reagan as Administrator of ADAMHA. He led the agency from July 1984 to June 1988. Reagan appointed him Deputy Assistant to the President for Drug Abuse Policy at the White House in 1988.

3 Reagan, "The President's News Conference."

4 Reagan, "Radio Address."

5 Carnevale and Murphy, "Matching Rhetoric to Dollars," 299–322.

6 Pollin and Durell, "Bill Pollin Era," 88–91.

7 Louise Richards, memorandum to William Spillane, December 23, 1980.

8 Louise Richards, memorandum to James Cooper and William Spillane, February 12, 1981.

9 Based on a March 30, 1981 memorandum written by Dr. Mitchell Balter, Chief of the Applied Therapeutics and Health Practices Program in NIMH, to the NIMH Reports Clearance Officer, in response to a request to review the OMB clearance package for the survey. Balter had been a principal advisor on the prescription drug section of the questionnaire in prior rounds of the survey. Comments and reactions to the memo by NIDA officials were handwritten on a copy of the memo.

10 Porter, Jacqueline, Minutes of April 23, 1981 Extramural Technical Review Committee for 1981 Household Survey on Drug Abuse. The Committee members were Mitchell Cohen, Gallup; Irving Lukoff, Columbia University; T. Alan Ramsey, Psychiatrist; Laure Sharp, Bureau of Social Science Research, Inc.; and Paul Uppal, Council on Children and Families.

11 Tourangeau, Rips, and Rasinski, *The Psychology of Survey Response*, 230–54.

12 Miller et al., *National Survey on Drug Abuse*, 1, 8.

13 Herb Abelson, memorandum for the file, July 20, 1982.

14 I was in the Statistical and Epidemiologic Analysis Branch, headed by Nick Kozel, who had been chief of the Forecasting Branch and moved to DESA when DAWN was transferred from DEA to NIDA. I had begun working at NIDA in 1980, as a mathematical statistician in DDID.

15 The National Survey contracts have always had strong restrictions on contractor use of the data, including public releases and secondary analyses not specified in the contract.

16 Several years later, MTF reports began showing two "any illicit" estimates, one with inhalants and one without.

17 From the mid-1970s through 1998, the two primary reports produced from each survey were Population Estimates and Main Findings. Population Estimates, which had been called Population Projections in early years, contained primarily a set of tables showing the estimated numbers of users of each drug,

by demographics, with 95 percent confidence intervals. Main Findings was a more comprehensive report on the full results of the survey. It included a narrative describing the findings, based on percentages of the population using each drug, and details on the survey methodology.

18 Voss and Clayton, "Stages in Involvement," 25–31.

19 The design effect quantifies the impact on the variance of the various features of a complex sample design as is used in NHSDA. The standard error is the square root of the variance. With a simple random sample, the variance of the estimate of a proportion such as the percent of the population using marijuana, would be estimated as p(1-p)/n, where p is the estimated proportion and n is the sample size. But with a complex sample, involving clustering, stratification, and differential sampling rates (e.g., due to oversampling), the variance is usually higher than if a simple random sample had been used. The factor by which the variance is higher is called the design effect, abbreviated as deff. The variance of a prevalence estimate is therefore correctly estimated as deff*p(1-p)/n. The deff can be different for different demographic groups and for different drug use measures, so it can be inaccurate to apply a constant deff for all of the estimates produced from a survey.

20 Charles Schuster became NIDA director in 1986, just prior to the October 9 press conference, which he attended, on the release of the 1985 NHSDA results.

21 Harrell, "The Validity of Self-Reported," 12–21.

22 A limited application of imputation was used by ISR for the 1985 data. However, it was only used in the Population Estimates report, and imputed values were not retained in the data file. A few years later under a SAMHSA contract, public use files were created for the 1979, 1982, and 1985 surveys, and new imputed variables were created and included on these files.

23 Kalton, "Compensating for Missing"; Groves et al. *Survey Methodology*, 329–33.

24 SAMHSA, *Results from the 2013*, 99–112.

25 Adams and Gfroerer, "Elevated Risk of Cocaine," 523–7; Adams, Rouse, and Gfroerer, "Populations at Risk," 25–41; Schober and Schade, *The Epidemiology of Cocaine*, 5–18, 253–62.

26 Johnston et al., "Use of Alcohol."

4 The White House Needs Data and a Bigger Survey

The Anti-Drug Abuse Act of 1988 had both immediate and long-term impacts on the drug survey. The law stated that ADAMHA "must collect data each year on the incidence and prevalence of the various forms of mental illness and substance abuse," for the nation and in major metropolitan areas selected by ADAMHA.

In addition, the act created the White House Office of National Drug Control Policy (ONDCP), headed by the drug czar. This also had major implications for the survey. The law gave ONDCP authority to establish policies and priorities for the National Drug Control Program, including coordination and oversight of agencies involved in drug-related activities, and the authority to review and approve the budgets of these agencies.

Another section of the law required ONDCP to develop and publish a drug strategy 180 days after the first director (the drug czar) was confirmed by the Senate, and by February 1 each year thereafter. The strategy was to include goals and measurable objectives. This stipulation guaranteed there would be a greater need for relevant, consistently collected data from the NHSDA and other sources in the coming years. Furthermore, federal agencies collecting that data would be required to provide their data to ONDCP, according to the law: "Upon request of the Director, and subject to laws governing disclosure of information, the head of each National Drug Control Program agency shall provide to the Director such information as may be required for drug control."

The bill was signed into law by President Reagan on November 18, 1988, two weeks after his vice president, George H. W. Bush, had been elected to succeed him in January. Once he took office, Bush nominated William H. Bennett to be the first director of ONDCP. Bennett had been Secretary of Education under President Reagan, and he promoted a hard-line moralistic approach to the drug problem. In his opinion, using drugs was wrong because drugs destroy a person's moral sense. He favored harsher penalties for drug users and dealers, and the use of the military in the war on drugs.[1] Bennett was sworn in as drug czar on March 9, 1989. He recruited two senior officials from the Department of

Education to help run ONDCP. John Walters, his aide on drug abuse issues at Education, became the deputy director of ONDCP, and Bruce Carnes came to ONDCP as Director of Planning, Budget, and Administration.

In the months following the passage of the Anti-Drug Abuse Act, there was a flurry of activity associated with planning for the 1990 and 1991 NHSDAs. The 1988 survey was unaffected since it was already in the field.

A National Drug Index

An influential leader during the Bush administration was James Burke. He had just retired after thirteen years as CEO of Johnson & Johnson Corporation. Lauded for his effective handling of the Tylenol poisoning scare in 1982, Burke became active in the war on drugs, and was appointed as Chairman of the Partnership for a Drug Free America in 1989. He was referred to as "the private sector drug czar."

In December 1988, just before the new administration took office, Burke met with President-elect Bush to tell him about how the private sector could help in the war on drugs. One of his recommendations was the establishment of a national drug index, to be published by the government every six months. Burke described it as "a 'tracking device' on our progress in reducing the consumption of drugs."[2]

After the inauguration, Burke met with NIDA to discuss how the survey could produce the index. I was asked to develop a plan. Borrowing some ideas from Richard Clayton and Harwin Voss,[3] I constructed a measure that was essentially a count of the number of drug use episodes in the population. It was the sum of the numbers of times people in the United States used each illicit drug within the past twelve months. A preliminary estimate for 1985 was 2.6 billion episodes. NIDA Director Schuster followed the meeting with a letter to Burke, in which he explained the details of a plan for the index, which would be produced every three months. However, Schuster's letter bluntly laid out a list of implications. We did not want to take on such a major task without sufficient staff and funding to support it. The index would require a larger, quarterly sample, questionnaire changes, more staff to manage the project, and increased budget to cover contract costs. Burke met with HHS Secretary Louis Sullivan to tell him of his work on drug abuse issues, and apparently to garner his support for expanding the NHSDA. A few weeks later, Burke received a letter from HHS Assistant Secretary for Health James Mason regarding the issue. Mason appreciated Burke's contributions and work with the Partnership for a Drug

Free America. He politely told Burke that due to budget constraints, the survey could not be conducted quarterly, but that NIDA was exploring the possibility of conducting the survey annually. Mason said "The scientific and social utility of a more frequent survey must be weighed against the personnel and monetary resources required for such a project."[4]

HHS Responds to the New Requirements

NIDA moved ahead with survey modifications according to the specific language on data collection in the Anti-Drug Abuse Act. Internal discussions led to a decision to expand the 1990 sample by specifying a sample size of 2,500 respondents in six metropolitan areas: Washington, DC, New York, Miami, Chicago, Los Angeles, and Denver. This would increase the overall sample size to about 20,000, and add about $6 million to the survey cost. Another $1.6 million was added to the contract to conduct a series of methodological studies during 1989 and 1990. This was an idea that I had proposed. With the coming expansion (metropolitan areas and possible annual data collection) and increasing reliance on the survey by policymakers, I felt that we needed to know more about how well the survey methods were working and whether there were better ways to collect the data. This work is described later in this chapter and in Chapter 5.

The $8 million expansion of the 1990 NHSDA – more than tripling the total cost of the 1988–90 contract – was processed as a sole source procurement to RTI, because the metropolitan area data collection and the methodological studies needed to be integrated with the national data collection effort. Having this work done under a separate contract would have resulted in inefficiencies, inconsistencies, and delays. The NHSDA team prepared a "Justification for Other Than Full and Open Competition," which required various approvals within HHS and a full review of RTI's proposal by an independent panel. This took several months. The contract modification was finalized on September 18, 1989. But while this process was moving forward, discussions within HHS on the plans for the surveys continued. At one point, NIDA was asked whether it would be feasible to conduct the 1990 survey in the spring so results could be ready in the same year. Our response was yes, this would be possible, but only if the six-city oversample was postponed until 1991. There simply was not enough lead time to select the additional sample, hire the needed staff, and complete 20,000 interviews in the spring. NIDA's recommendation was to stick with the current plan of a fall data collection with the metropolitan oversampling.

There were other unresolved questions. Did the survey really need to be done every year? How often should metropolitan areas be oversampled? Should the same six remain or should other areas be rotated in? Options, with pros and cons, were sent to ADAMHA Administrator Dr. Frederick Goodwin, who sent them on to Assistant Secretary for Health James Mason to make the final decision.

NIDA first learned of Mason's decision in a phone call on September 26. He rejected NIDA's recommendation. He said the 1990 survey must be done in the spring, with results available as soon as possible in 1990. Now we had to undo the sample design plan specified in the $8 million sole-source contract that had taken half a year to process, and had been signed just one week earlier. The contract for the 1990 survey was modified again, shifting data collection to the spring and dropping five of the six metropolitan area oversamples. Based on our discussions with the contractor, we determined it would be feasible to keep the Washington, DC oversample without jeopardizing the new schedule. The Washington oversample was important because of its contribution in another NIDA study, also done by RTI under a separate contract, called the DC Metropolitan Area Drug Study (DC-MADS). This study was an attempt to capture the prevalence of drug use across all populations in a single metropolitan area by surveying the homeless, institutionalized, and other special populations, and combining the data with the NHSDA sample to result in a more complete picture of drug abuse in the area. It was considered a methodological study – a feasibility test.[5]

Other methodological studies that had been specified in the sole-source NHSDA contract and funded for $1.6 million, were retained in the modified 1990 survey plan. The primary study in this set was (1) a 2,000-respondent field test that would experimentally test different data collection approaches and estimate their effects on drug use estimates and data quality. Other studies included (2) cognitive evaluations of the questionnaire to understand how respondents understand and perceive the wording of key questions; (3) analysis of patterns of inconsistent and invalid responses to specific questionnaire items; (4) detailed analysis of non-response patterns in the 1988 NHSDA; (5) a non-respondent follow-up study that would obtain data from sampled persons who do not participate in the survey, to better understand the bias in NHSDA estimates caused by the approximately 20 percent of sample persons for whom no data is obtained; (6) Census match study, which would link non-respondents in the 1990 NHSDA to their 1990 Decennial Census data, to gain understanding about non-respondents; and (7) a study exploring and assessing different sample design options, in terms of costs, data quality, and sampling error. Results from these studies are summarized in Chapter 5.

Mason's formal response to the decision memorandum on the NHSDA design options came on October 16, 1989: "I believe it is critical that data from the basic 1990 National Household Survey be available no later than the end of 1990, earlier if possible." The memo also instructed NIDA to "incorporate the metropolitan area oversamples in the spring 1991 and subsequent annual surveys."[6] No further explanation for the decision was provided, but the public and political pressure to provide the data ONDCP needed for the annual drug control strategy was intense. President Bush had promised that "This scourge will end." Drug abuse policy was prominent in the news and the drug problem was a top concern among Americans during this era. HHS had little choice.

With the 1990 plans seemingly finalized, the NHSDA team turned to the design for the 1991 survey. The survey would be done in the spring of 1991, with the six-city oversample included, for a total sample size of about 20,000. The contract was awarded several months later to RTI, but soon there would be more changes to the sample design.

ONDCP Gets Involved

Just six weeks after he was confirmed by the Senate to be the drug czar, William Bennett sent an ominous memo to HHS Secretary Louis Sullivan and other agency leaders.[7] The memo stated that all drug-related materials must be submitted to ONDCP for review prior to release. This included reports, studies, evaluations, proposed regulations, testimony, and other items. In addition, materials submitted by agencies to OMB for clearance would be reviewed by ONDCP, during the clearance process. This meant that every questionnaire, every sample design plan, and every report of the results from the survey would be reviewed by ONDCP. ONDCP would have an opportunity to request (or demand?) changes to any of these materials. NIDA officials were concerned that political motivations might now trump science, and the extra level of review could cause delays.

As a new agency, ONDCP needed to quickly build a staff to handle all of the responsibilities they were given. Besides reviewing all drug-related documents across the entire federal government, they had to prepare the interim National Drug Control Strategy by September, and then an annual strategy every February that reported on progress toward achieving measurable goals. Data would be needed for these and other reports. They needed data on the prevalence of drug abuse, primarily from the NHSDA and the MTF. They also needed data on persons receiving treatment, the characteristics of treatment programs, health consequences of drug abuse, criminal activity related to drugs, production and supply data from countries all over the world, price and purity of drugs, and other data, from dozens of sources. They also had to identify data gaps, and decide how data systems should be revised to meet the needs of the drug war. Inevitably, they would need to rely heavily on the experts present in each of the federal drug control agencies.

In mid-1989, ONDCP analysts began contacting me and other members of the NHSDA team. They asked questions about the data sets (primarily NHSDA and DAWN), inquired about our plans, and requested special analyses. At first we were glad to help, but it soon began to interfere with our regular duties managing the survey and producing NIDA reports. If we stopped those critical activities to work on ONDCP requests, we would put future data collection schedules and reports at risk. The ONDCP analysts were not sympathetic. In their minds, White House requests should be a priority.[8] After we explained the situation to ADAMHA leadership, the agency made efforts to divert the requests elsewhere. First they required all inquiries to be routed though a point of contact in the administrator's office and then through a point of contact in NIDA. That system was not satisfactory to ONDCP, so NIDA eventually (in 1991) agreed to create a dedicated team to handle ONDCP requests for analysis and quick response surveys. This team was called the Special Projects Group (SPG). After a hasty set up, the SPG was not able to fully satisfy ONDCP needs.[9] The SPG staff simply did not have the expertise and knowledge of NHSDA and other data sets, or the analytic skills needed to quickly develop the studies that ONDCP requested. Physically situated near the NHSDA team, SPG staff often approached my staff to ask for help on their ONDCP assignments. So the SPG did not fully accomplish its goal of diverting the ONDCP requests from the NHSDA team. The intimate knowledge of the data that staff develop through years of working directly on a survey can rarely be matched by separate analytic units like SPG. Hands-on involvement and responsibility for developing the questionnaire, data collection, editing, imputation, weighting, and file construction by staff builds a knowledge base that is critical to the success of an ongoing large survey, especially in analysis, report writing, and responding to data requests.

The ONDCP requests for data analysis or quick-response surveys were sometimes unclear. To design and conduct the most relevant analysis or survey, NIDA often replied to ONDCP's requests by asking them to be more specific and explain the goals of the study. Knowing the capabilities and caveats of the data sets we had, we were in the best position to decide exactly how to design the study so that the results would meet ONDCP's goals. But to ONDCP, these replies were intrusive and conveyed that NIDA did not trust them.

The First National Drug Control Strategy

ONDCP's frequent requests for help obtaining and analyzing NHSDA data were mainly from ONDCP staff working on the National Drug Control Strategy. Bill Bennett became drug czar in March 1989 and had six months to produce the interim strategy. The strategy was written based on the belief that the drug problem was getting worse. The violence associated with the crack epidemic was increasing, cocaine seizures were up, and emergency department episodes involving cocaine and heroin were increasing. Unfortunately, the most recent NHSDA data available when work began on the strategy was the 1985 data. By the time the 1988 estimates were available to ONDCP in July 1989, the strategy was nearly completed. In contrast with the theme of the strategy, the survey results showed a substantial decline in marijuana and cocaine use. This caused disbelief and confusion at ONDCP. The reaction of ONDCP leadership was to question the survey and to discount its findings. They incorporated the new NSDUH results into the strategy document, but retained the emphasis on the worsening of the drug problem.

The interim National Drug Control Strategy was released by ONDCP in September 1989, in conjunction with a televised speech by President Bush. He declared an escalation of the war on drugs, with increased funding for jails, prisons, courts, and prosecutors. To vividly show how widespread America's drug problem was, he held up a bag of crack cocaine that he said undercover agents recently bought in Lafayette Park, across the street from the White House. It was later revealed that federal agents had lured someone to the park to sell crack just so the president could say it was bought in front of the White House.[10] What this stunt really demonstrated was the penchant of some politicians and government leaders to distort and even invent facts to make a point.

The opening paragraphs of the strategy were telling:

> In late July of this year, the Federal Government's National Institute on Drug Abuse (NIDA) released the results of its ninth periodic National Household Survey on Drug Abuse – the first such comprehensive, national study of drug use patterns since 1985. Much of the news in NIDA's report was dramatic and startling. The estimated number of Americans using *any* illegal drug on a "current" basis has dropped 37 percent: from 23 million in 1985 to 14.5

million last year. Current use of the two most common illegal substances – marijuana and cocaine – is down 36 and 48 percent respectively.

This is all good news – very good news. But it is also, at first glance, difficult to square with commonsense perceptions. Most Americans remain fully convinced that drugs represent the gravest present threat to our national well-being – and with good reason. Because a wealth of other, up-to-date evidence suggests that our drug problem is getting worse, not better.

With this introduction, ONDCP simultaneously acknowledged the central role NHSDA would play in their assessment of the drug problem, and dismissed the survey's findings as not accurately portraying the scope of the drug problem. Repeating the drug czar's speech at the July press conference (see Chapter 3), the interim strategy claimed "What, then, accounts for the intensifying drug-related chaos that we see every day in our newspapers and on television? One word explains much of it. That word is crack."[11] The strategy described a two-tier drug war. While casual use was declining, heavy or "hard-core" use was increasing. ONDCP cited the twenty-eight-fold increase in emergency department episodes involving smoked cocaine since 1984.[12] Aware that the 1985 and 1988 NHSDA estimates of frequent cocaine use (weekly or more often) were not significantly different, ONDCP focused on a different statistic, that did show a significant increase. They cited the estimate of frequent cocaine use *calculated as a percentage of past year cocaine users*. That estimate doubled from 5 to 10 percent between 1985 and 1988.

Later in the strategy report, in describing a research agenda, ONDCP discussed weaknesses of the NHSDA: the survey was done too infrequently (every three years), the sample size was too small, the data on intravenous and heroin use were poor, and the data relied on self-reports. These deficiencies had become clear to ONDCP during the preparation of the strategy. To partially address these weaknesses, ONDCP stated:

The Administration has committed to funding more frequent National Household Surveys – every two years. The Federal Government will also create a quick response capability to permit smaller, narrowly targeted surveys undertaken several times a year. Such surveys will enable us to focus on particular groups in the population – "high risk" youth, for example – or on emerging drug trends.

Of course, the biennial data collection had already been established with the upcoming 1990 survey. That decision had been made two years earlier. In fact, HHS decided a month after the interim strategy was released to conduct the survey *every* year. The quick response survey was an idea that Bennett and the staff he brought from the Department of

Education to ONDCP were familiar with. Education had set up a quick-response survey capability. But in the world of OMB clearances, Institutional Review Boards, review processes across the bureaucracy, and an ineffective Special Projects Group, the quick response survey capability was never fully developed.

ONDCP Tells NIDA to Increase the Sample Size

Not long after the release of the interim strategy, ONDCP began to exert their influence on the NHSDA design. NIDA had submitted the request to OMB for clearance to conduct the 1990 NHSDA, with a national sample of 8,000 plus an added oversample in the Washington, DC metropolitan area. A conference call took place in late December to talk about ONDCP's review of the 1990 survey design described in the OMB request. Leading the discussion for ONDCP was Ms. Gabrielle Lupo, Director of Planning, under Bruce Carnes. The discussion focused on ONDCP's desire for a larger NHSDA sample. They wanted estimates with better precision and a larger pool of drug users for special analyses. Gabi asked if the sample size for the 1990 survey could be doubled. With the data collection starting in just three months, and the initial stages of the sample already selected, we told her it was probably not possible to double the sample at this late date. She was insistent, so we agreed to discuss the possibility with RTI. Gabi also inquired if there was a way we could incorporate a revised screening procedure into the survey to identify households with drug users. We said we would also explore this idea.

We contacted RTI and discussed the proposed changes. RTI said they could double the sample. But it could only be done at the final stage of selection – by selecting more people in the sample of households that had already been selected. There was not enough time to select and contact a large number of additional addresses. But adding more respondents without increasing the number of areas or households would result in a relatively small improvement in the precision of estimates. Also, we were concerned about increasing the workload for field staff on such short notice, when it was critical that results be available in December 1990. More field staff would need to be hired, quickly. As contractors sometimes do when faced with funded expansions to the scope of work, RTI said they were capable of pulling off this major increase to the contract. But our perspective was that this would be the largest survey they had ever conducted, and data collection would begin in three months. Results were needed by the end of the year. Despite RTI's assurance that this work could be completed, NIDA could not take the risk, and told ONDCP it could not be done.

The 1990 National Drug Control Strategy was released on January 25, and this time the research agenda was more ambitious than it had been in the interim strategy published four months earlier. The new direction for the survey was:

> The National Household Survey on Drug Abuse will be conducted annually, rather than every three years, and the sample size will be increased from 8,800 to 20,000 households. Changes will be made in the survey content to reflect the need for information more directly relevant to drug policy as well as epidemiological concerns. Data from the first of these revised surveys will be available for use in the development of the next National Drug Control Strategy.

Other new data collection was specified in the Strategy. ADAMHA was to begin "Quick Response and Target Surveys," with the first data to be ready by January 1991. The Monitoring the Future study of high school seniors was to be expanded to cover earlier grade levels, as well as school dropouts. ADAMHA was to expand prevention research. More detailed data on the drug abuse treatment system would be obtained through NIDA's National Drug and Alcoholism Treatment Unit Survey, including follow-up studies of patients after treatment, to study treatment effectiveness.

As required by the Anti-Drug Abuse Act, the strategy included a set of quantified objectives. For this initial strategy, there were ten two-year objectives and ten ten-year objectives. They included measures of use, consequences, drug availability, production, and attitudes. Ten of the twenty objectives were based on NHSDA data. For example, objectives for current overall illicit drug use were a 15 percent reduction within two years and a 55 percent reduction within ten years. Upon review, we pointed out that because of its small sample size (9,259 respondents) the 1990 survey would not have enough precision to accurately track progress for some of these objectives. For example, one objective was to reduce adolescent cocaine use by 30 percent in two years. But a 30 percent decline would not reach statistical significance, based on the estimated sampling errors for the 1990 sample design. Between 1985 and 1988, the estimate of current cocaine use among youths did decline about 30 percent, as did hallucinogen use. Inhalant use declined about 40 percent. None of these "declines" were statistically significant, meaning there was not enough evidence to state with confidence that a decrease actually occurred. The larger samples beginning in 1991 would improve the tracking capability, but even with the 1991 data, the rate of decline could only be based on a comparison with the smaller 1990 sample. Accurate assessment of change over a two-year period would not be possible until the 1993 estimate was compared with the 1991 estimate.

ONDCP's insistence on the sample increase for 1991 was encountered again through the OMB clearance process. Midway through February 1990, we still had not received approval to conduct the 1990 survey. If approval did not come soon, we might have been forced to postpone the survey, jeopardizing our commitment to have the results ready in December. We were also concerned that delays would increase costs. Training and data collection arrangements might have had to be canceled and rescheduled, and staff would have to work overtime to complete the work on a compressed schedule. OMB finally cleared the survey on February 20, but ONDCP had inserted conditions on the approval. They demanded that NIDA modify the 1991 survey as they had stated in the Drug Control Strategy.[13] The 1990 survey was approved only if the 1991 national sample was "increased to the equivalent of 20,000 households sampled on a simple random basis representative of the US population." They also wanted questionnaire changes, including "general questions to screen for overall drug use," questions on attitudes about drug use, and new questions on health insurance coverage and income.

NIDA Responds to ONDCP's Demands

In an effort to keep ONDCP informed and involved in survey design planning, and to avoid delays in survey design decisions, on February 26, 1990 NIDA sent a memo to ONDCP saying that the base national sample for the 1991 survey would be increased to 20,000, with additional oversamples in six metropolitan areas.[14] The contract for the 1992, 1993, and 1994 surveys was to be awarded in March, and it was important to have final decisions on the design for those surveys prior to awarding the contract. The memo also described plans for methodological studies. The last paragraph stated that NIDA's Division of Epidemiology and Prevention Research was "very much interested in working with you to make the future National Household Surveys responsive to all our needs ... it is essential that we begin discussions as soon as possible in order to avoid disruptions in the data collection schedule." Despite the gesture, the relationship between ONDCP and NIDA would continue to deteriorate over the next several months.

Although it may have appeared as if NIDA was proceeding according to ONDCP's wishes, NIDA was unhappy with ONDCP's meddling. NIDA leaders opposed the expansion of the NHSDA, and they were not ready to give in. Consistent with the opinions of its leading grantees, NIDA felt that the additional funding for drug abuse epidemiology should be used to expand research grants, not large national survey contracts. NIDA Director Schuster convened a "Research Retreat" in

Annapolis, MD from March 13–15, 1990, to discuss the ONDCP demands and solicit opinions on the future direction of NIDA's epidemiology program from a group of the nation's top substance abuse researchers.[15] Several were principal investigators on NIDA-funded grants. Despite the possible conflict of interest, when faced with important decisions concerning data and surveys, NIDA directors often turned to these colleagues for advice. Also in attendance were key staff from NIDA's DEPR.[16] I do not recall anyone from ONDCP attending the meeting.

During the meeting, DEPR staff described the current epidemiology program and the changes that ONDCP wanted, along with concerns about the impact on trend integrity. I pointed out that NHSDA expansion would require a large field staff increase, resulting in many new and inexperienced interviewers and changes in management of data collection. These changes could potentially affect response rates and respondent reporting behavior, which could affect estimates. The consultants generally agreed. They urged that the survey remain biennial with a sample of about 8,000, and that additional data needed by ONDCP should be obtained through separate studies done by grantees. They also said there was a continuing need for methodological studies to increase understanding of the survey data, and that these studies often can be conducted best as a part of "substantive research by leading researchers in the field, instead of vendors who happen to be available at the time an RFP for a contract is issued." This comment reflected a pervasive theme throughout the meeting. The consultants were clearly concerned that the recent legislation and ONDCP demands were going to result in funding increases to "beltway bandits" (contractors) instead of the existing army of university-based researchers (grantees). They said "The basic surveillance reporting activities could be handled under the contract mechanism," while the "higher level epidemiologic analyses" could be done through grants. They complained that during a recent peer review of grant proposals, NIDA officials said there was no money to fund any of the grants, while at the same meeting, the review panel was told that DEPR was initiating new contracts and was recruiting to fill epidemiologist positions. Ironically, the consultants suggested that a peer review process for the "quasi-intramural activities of the Division" was needed, to eliminate the appearance of a conflict of interest between extramural and intramural programs. While some of these concerns by the outside consultants had merit, the DEPR staff in attendance were offended by the self-serving criticisms. The final hour of the retreat was even more disconcerting. To our surprise and dismay, Schuster asked all of the

DEPR staff to leave the room, so he could close out the retreat by talking privately with the outside consultants.

A draft summary of the meeting was sent to Schuster on March 28.[17] The first paragraph of the summary seemed to indicate that these consultants may have misunderstood NIDA's predicament, and overestimated their own potential influence, relative to that of Congress and ONDCP:

> The National Household Survey on Drug Abuse is a specific topic that requires immediate attention and action. Two changes in the work-scope of the National Household Survey are being introduced, or are proposed for introduction soon, apparently with active encouragement from officials at ONDCP.

Perhaps Schuster knew ONDCP's demands were more than "active encouragement." ONDCP had the authority to force the NHSDA expansion. The retreat may have truly been a last-ditch effort by Schuster to derail the ONDCP directives. Or maybe his underlying purpose for the meeting was to give these highly valued NIDA consultants the impression that their opinions mattered, and that the report on the research retreat would influence the decision-makers. It did not. On March 29, the day after NIDA received the draft meeting summary, officials from ONDCP, HHS, ADAMHA, and NIDA met at ADAMHA's headquarters in Rockville, MD, to discuss NHSDA issues. At the meeting, NIDA agreed (again) to increase the national sample component of the 1991 and subsequent annual National Surveys to the equivalent of 20,000 respondents. NIDA's plans for methodological studies were also discussed at the meeting. ONDCP expressed great interest in these studies, especially the field test. They anticipated that the field test would yield a method to screen for and oversample drug users, resulting in a powerful data set to use in studies of hard-core drug users.

A few weeks after the Rockville meeting, NIDA sent another memo to ONDCP, updating them on upcoming NIDA studies, and inviting them to tell us about any new requirements they were considering.[18] Once again, the purpose of the memo was to avoid last-minute design changes or delays in approvals, especially for studies involving data collection that require OMB approval. But over the next several months, methodological studies and the 1991 survey were jeopardized due to delays in OMB approval caused by ONDCP involvement. All of the studies were eventually approved, but the delays resulted in added costs. And each approval came with conditions. This was ONDCP's mechanism for imposing their will on NIDA data systems. OMB clearance requests were held hostage until ONDCP was satisfied NIDA had sufficiently acceded to their demands.[19]

Addressing Survey Methodology Concerns

ONDCP wanted us to find ways to oversample drug users, and to measure and adjust for nonresponse. Fortunately, we already had in place a $1.6 million program of methodological research, in the 1990 NHSDA contract. To us, oversampling drug users translated to a stratified design that increased selection probabilities in populations where we believed there was an elevated rate of drug use. However, ONDCP at first was expecting that we could actually screen for drug users, such as including a question on drug use in the household screener: "Does anyone in the household have a problem with drug use?" After we pointed out the potentially negative effects of such a procedure, they suggested simply inserting skip patterns within the questionnaire, to ask early during the interview if a respondent had used a drug, and then having those who report no use skip the remaining questions about the use of that drug. We agreed to study that approach, but there was concern that skip patterns would lead to underreporting. RTI developed comprehensive discussion papers that provided some ideas of how the oversampling could be done. These were sent to ONDCP, but still did not satisfy them. They wanted skip patterns inserted in the questionnaire as soon as possible.

ONDCP's request for studies of nonresponse could be answered without much additional effort. We simply compiled descriptions of the several nonresponse-related studies we had planned. But when we submitted them to ONDCP, the reply was that the studies did not address their concerns. We were puzzled. After several attempts to resolve the quandary, we finally discovered a simple communication problem. To ONDCP, nonresponse referred to the failure of some respondents to report their drug use. In survey methodology though, that underreporting is referred to as "measurement error" or "response bias."[20] Nonresponse occurs when data are missing because respondents either don't participate (unit nonresponse) or fail to answer a particular question (item nonresponse). The misunderstanding resulted in months of delay, wasted effort by NIDA and RTI staff, and extra costs to the government.

Another survey design concern we began to explore after ONDCP had demanded the expansion of the national sample was the shifting data collection period of the survey. This could potentially affect comparability across years. Past surveys had been relatively small, and were conducted during a period of a few months – a different time of year for each survey. However, with an annual sample of 30,000 respondents, a better approach was to spread the interviewing over the entire twelve months of each year, constituting continuous data collection. We proposed this for the next contract, covering the 1992, 1993, and 1994 surveys. It had many advantages. The data collection would be done with a smaller, more committed field staff, resulting in better data quality and potentially lower costs. Seasonality would no longer be a concern in comparisons of estimates across years. Analytically, we would have a data set better able to track trends in drug use and associate them with different policy changes or events that occur on specific dates. We decided the best way to spread the interviews uniformly throughout the year would be to select a new sample each quarter. Interviewers would have three months to complete their assignments, and there would be no spillover into the next quarter. Workloads would be such that it would take about two months to finish most of the work, with the third month of each quarter reserved for "clean-up" of difficult or lagging cases. The survey could produce nationally representative estimates for each quarter. This sampling plan would facilitate Jim Burke's idea for a drug index. ONDCP liked the idea, as it had the potential to give them more timely data. NIDA Director Schuster also supported this new design, but cautiously limited it to the 1992 and 1993 surveys, to allow greater flexibility to change the sample design in 1994. Finally, in December 1990, Assistant Secretary for Health Mason approved. Shortly thereafter, the contract for the 1992 and 1993 surveys was awarded. Once again, RTI was the winning bidder.

During this time of important decision-making about the NHSDA design, NIDA continued to get outside opinions and discuss them with ONDCP. We convened a group of consultants on April 18, 1991 to discuss sample design options, with a focus on the goal of oversampling drug users. The consultants recommended that any further design changes should be planned for 1993 or later, to allow time for development and testing.[21]

A "Methodology Workshop" was held on April 19, 1991. A panel of survey design experts and drug abuse researchers reviewed and discussed early results from our 1989–1990 methodological studies.[22] Some of their recommendations on the NHSDA survey design were:

- Continue to use a self-administered questionnaire. Due to the increasing sensitivity associated with smoking, consider making that section also self-administered.
- Use respondent-implemented skip instructions.

- Continue to use the calendar and questioning strategy that emphasizes the boundaries of the reference periods.
- During training, provide the data collection staff with information on the results from the field test so that they can see the value of some of the procedures that are more difficult to implement.
- Investigate the use of self-administered interviews that are computer assisted to permit automatic branching, screening for drug users, and privacy of response. In the long term, consider developing a computer assisted interview that makes use of voice recordings for those who have difficulty reading.
- Continue to investigate ways for reducing the inconsistencies in the response focusing on understanding of reference periods, concepts, and answer categories.
- Focus on ways to reduce overall nonresponse. Consider testing an incentive on a subsample of the main survey.

As we were preparing for the 1992 and 1993 surveys, NIDA needed to decide which metropolitan areas to oversample each year. NIDA Director Schuster wanted areas different from the ones oversampled in 1991. He also wanted to choose from the set of twenty-one areas that were oversampled in the DAWN, so data from the two surveys could be analyzed together. In January 1991, DEPR proposed Detroit, Boston, Baltimore, Atlanta, San Francisco, and San Diego to Schuster. He concurred. But in May, Jim Burke from the Partnership for a Drug Free America met with us and indicated that NIDA should consider keeping the same metropolitan areas in NHSDA each year. He reasoned that this would facilitate assessing the impact of prevention efforts in those original six cities. A few weeks later Schuster decided we should follow Burke's recommendation. ONDCP did not object. Their interest was shifting towards state rather than city data.

ONDCP Pushes for State-Level Estimates

The communication difficulties between ONDCP and NIDA continued in April 1991, just four weeks after Bob Martinez, former Florida governor, was sworn in as the new drug czar. On April 25, HHS Counsel for Drug Abuse Policy Mark Barnes received a letter from Bruce Carnes, Director of the ONDCP Office of Planning, Budget, and Administration.[23] "I am extremely concerned to learn that NIDA is formulating the sample design and conduct of the 1992 survey without having obtained policy guidance. We would like to see state and not city specific data, and a complete reworking of the survey content." DEPR staff was shocked by this unexpected memo from Carnes, coming only one week after we had met with Gabi Lupo to talk about the sample design, and only eight months before the survey would go into the field. We promptly put together documentation of all of the communication difficulties. A memo from ADAMHA to the Assistant Secretary for Health was prepared. The memo described our attempts to involve ONDCP in design decisions, including three face-to-face meetings between ONDCP and NIDA in the past two months, at which the sample and questionnaire changes described in Carnes' memo had not been mentioned by ONDCP representatives. Over the past year, NIDA had made numerous attempts to work with ONDCP, but "It has become increasingly clear in our frequent dealings with ONDCP on data issues that the staff there have great difficulty in articulating what research questions they want answered by these data systems. They also seem to lack a basic understanding about the nature of the data systems they continue to try to modify ... The absence of survey research expertise at ONDCP underlies the problems we have experienced." The memo went on to state "I do not appreciate their uninformed efforts to micromanage NIDA's epidemiologic research program ... ONDCP appears to be usurping the role of the Secretary in determining the type of information to collect regarding the health of the nation." In closing, the memo said "We cannot and will not sacrifice the integrity of our internationally recognized epidemiologic research program by being party to poorly articulated, not well conceived ONDCP insights ... Until such time as ONDCP can clearly define its needs in accurate terms, we are unable to respond with a strategy for responding to their April 25 request."

A few weeks later, Barnes (HHS) met with Carnes (ONDCP) and explained why the Household Survey was not the best vehicle for collecting state-specific data. Carnes responded by saying he would continue to hold up clearance packages until HHS provided some alternative ways of collecting state data on drug abuse. Shortly after that, Gabi Lupo sent a terse note to Mark Barnes: "You asked that I provide a specific statement of what ONDCP is looking for. O.K. It is, 'State-specific data on prevalence for major drugs of abuse and major population and age groups.'" This was ONDCP's guidance to NIDA on the further expansion of NHSDA to get state estimates, an expansion estimated to cost as much as $36 million per year.

In June, ONDCP finally articulated the changes they wanted to the questionnaire before they would approve the 1992 survey. Bruce Carnes' memo to Mark Barnes indicated "In my view, these do not constitute a major disruption to the 1992 Survey plan or schedule, and I recommend that NIDA make these changes."[24] Actually, they were very significant changes that would require considerable time to develop and test. They included inserting skip patterns, dropping questions on misuse of prescription drugs ("These do not add to the data on illegal drug use."), changing interview procedures and definitions, and adding new questions. Carnes also included a list of "more ambitious" changes for the 1993 survey, saying he "would like to give you and NIDA as much lead time as possible." These ideas included: (1) oversample all drug users; (2) move income questions to the beginning of the questionnaire; (3) group drug questions by time period; and (4) rotate different modules in and out of the questionnaire throughout the year.

We prepared a four-page response explaining what we could do and when, and why many of the proposed changes were problematic.[25] Skip patterns could inhibit respondents' reporting of drug use. Prescription drug abuse was an important public health concern that needed to be tracked. Oversampling all drug users was not feasible. We should not move the income questions, as it's well known among survey researchers that income is a highly sensitive topic that is best asked near the end of an interview. Grouping questions by time period and rotating modules would require a huge effort to develop and implement. Our response satisfied ONDCP, with a couple of exceptions. They strongly felt that all questions should be consistent across drug types, and that a split sample should be employed in 1992 to measure the effect of the new, consistent wordings.[26] We resisted, because we needed more time to develop and test these major questionnaire changes. Eventually, an agreement was reached and ONDCP allowed the 1992 OMB approval process to proceed without further interference.

There was progress on the state data issue as well. Mark Barnes convened a meeting on August 7 with representatives from ONDCP and HHS to discuss state data collection efforts and initiatives. ONDCP and HHS agreed not to pursue the NHSDA as a means of collecting state data, at this time. Bruce Carnes discussed the administration's policy regarding closer monitoring of the states' use of Block Grant funds and the important role that states play in providing drug abuse treatment and prevention services. ADAMHA's Office for Treatment Improvement made a presentation of their State Systems Development Program (SSDP), which was in the planning stages at this point. SSDP would provide funds and technical assistance to states to conduct their own drug use surveys, resulting in state and substate level estimates of drug abuse treatment need. The surveys would be conducted by telephone. Carnes thought the SSDP was impressive, and indicated that ONDCP would support this effort. NHSDA would remain focused on national and metropolitan area estimates, for now.[27] But ONDCP still viewed the

Household Survey as the core data source for assessing national trends and patterns. Their desire to expand the NHSDA to provide state-level data persisted.[28] ONDCP needed data to better understand how state efforts, particularly those supported by federal formula and competitive drug control grants, were performing.

Notes

1 Bennett, "Penalties Must Be Harsh."
2 J. E. Burke, letter to President-elect George Bush, December 13, 1988.
3 Clayton and Voss, "A Composite Index."
4 James Mason, letter to James Burke, June 1990.
5 Bray and Marsden, *Drug Use in Metropolitan America.*
6 James Mason, memo to Administrator, ADAMHA, October 16, 1989.
7 William J. Bennett, memo to Louis W. Sullivan, April 21, 1989.
8 Eventually, ONDCP hired a contractor to conduct some of the analyses they needed. We provided NHSDA with microdata files for them to use. This arrangement gave ONDCP better control over the direction of the studies.
9 GAO, *Drug Control: Reauthorization,* 59–60.
10 Isikoff. "Drug Buy Set Up For Bush Speech: *DEA Lured Seller to Lafayette Park.* " *Washington Post,* September 22, 1989; Page A01.
11 Office of National Drug Control Policy. *National Drug Control Strategy, September 1989.* Page 3.
12 Data from the Drug Abuse Warning Network (DAWN).
13 The wording used by ONDCP in the strategy and in the OMB clearance conditions was confusing and imprecise and was not taken literally by NIDA. We assumed the 8,800 they referred to was the number of persons participating in the 1988 survey, not the number of households, which was more than 30,000. Their reference to a sample equivalent to a simple random sample of 20,000 was complicated by the oversampling of young people, racial/ethnic groups, and metropolitan areas in the current design. We decided to expand the sample to achieve precision equivalent to a sample of 20,000 respondents under a design with the same proportions allocated to age and race/ethnic groups, supplemented by the extra sample for the six metropolitan areas.
14 Edgar Adams, memorandum to Gabrielle Lupo, February 26, 1990.
15 Outside consultants in attendance included James Anthony, Dale Chitwood, Richard Clayton, Chris Hartel, Denise Kandel, Sheppard Kellam, Lloyd Johnston, Ben Locke, Roy Pickens, and Bruce Rounsaville.
16 DEPR staff included Edgar Adams, Ann Blanken, Joe Gfroerer, Nick Kozel, Susan Schober, and Betsy Slay.
17 Draft report on actions at the Annapolis Waterfront Hotel retreat. Prepared by James C. Anthony from notes by Kenneth R. Petronis, March 28, 1990.
18 Memo from Edgar Adams to Gabrielle Lupo, May 14, 1990.
19 ONDCP used the OMB clearance process to force agencies to make changes because the Anti-Drug Abuse Act gave them limited oversight authority over federal agencies. See GAO, *Drug Control: Reauthorization,* 53.

20 Groves et al., *Survey Methodology*, 49–56.

21 Attendees included Edgar Adams, Ann Blanken, Joe Gfroerer, Joe Gustin, Arthur Hughes, and Lana Harrison from DEPR; Gabi Lupo (ONDCP); James Anthony (Johns Hopkins University), Ralph Folsom (RTI), and Steven Botman (NCHS).

22 Expert panel consisted of James Anthony (Johns Hopkins University), Charles Cannell (University of Michigan), Richard Clayton (University of Kentucky), Tom Mangione (University of Massachusetts Boston), Nancy Mathiowetz (Bureau of the Census), Andrew White (National Center for Health Statistics), Gordon Willis (NCHS), and Ron Wilson (National Center for Health Statistics). RTI attendees were Tom Virag (Project Director), Valley Rachal (Co-Principal Investigator), Ralph Folsom (Co-Principal Investigator), Paul Moore (Methodology Studies Manager), Judith Lessler (Field Test Analysis Leader), Charles Turner (Field Test Analysis Team), and Lynn Guess (Analyst). Gabi Lupo from ONDCP attended, and various DEPR staff were present.

23 Bruce Carnes, memorandum to Mark Barnes, April 25, 1991.

24 Bruce Carnes, memorandum to Mark Barnes, June 20, 1991.

25 Mark Barnes, memorandum to Bruce Carnes, July 10, 1991.

26 Bruce Carnes, memorandum to Mark Barnes, July 22, 1991.

27 Mary Knipmeyer, Deputy Associate Administrator for Policy Coordination, ADAMHA, Summary of August 7 meeting with ONDCP regarding state data collection, August 28, 1991.

28 Gabi Lupo was detailed to NIDA (not DEPR) at the end of 1991, and was replaced as head of ONDCP planning by Dr. John Carnevale, an economist in another ONDCP unit under Bruce Carnes. Carnes asked Carnevale to improve relations with HHS, which he did. Carnevale then took over as Director of the Office of Planning, Budget, Research and Evaluation when Bruce Carnes left a year later. Carnevale held the position until 2000.

5 Criticism, Correction, and Communication

Soon after ONDCP announced the NHSDA would be a principal data source for tracking progress in the war on drugs, criticisms of the survey and alternative estimates began to appear. As the 1992 presidential election approached, the political rhetoric surrounding the drug war and the data used to evaluate it increased. The scrutiny intensified in 1992 when errors in the 1991 NHSDA estimates were discovered, and the General Accounting Office (GAO) conducted a study of NHSDA and other drug abuse data sources.[1] NIDA's comprehensive methodological research program to assess the NHSDA methods enabled NIDA to respond to criticisms and provided a basis for improving the survey.

Other Studies Challenge NHSDA Estimates

On October 1, 1989, the Parents' Resources Institute for Drug Education (PRIDE) released the results of their survey of students in grades six through twelve.[2] PRIDE officials said their survey, administered to nearly 400,000 students in 958 schools, found youth drug use was higher than NHSDA had shown. They claimed NIDA's survey understated the nation's drug problem, raising questions about goals established by the Bush administration. They explained that youths are reluctant to report their drug use in a survey conducted at home, where parents may be present. Although the comparison between PRIDE's sixth to twelfth graders and NHSDA's 12–17-year-olds was an apples versus oranges comparison,[3] the criticism was valid, based on studies of NHSDA and MTF data.[4] The student survey became a regular product of PRIDE in subsequent years, touted as the nation's largest youth drug survey. But despite its size, the PRIDE sample was a convenience sample, lacking any random selection of schools that would justify projecting the data to national estimates. The data came from schools voluntarily participating in PRIDE's data gathering service, which included the use of their questionnaire and school-specific tabulations. PRIDE combined all of the schools' data and announced the results as a "nationwide survey." An analysis of this sample by state showed that school participation rates varied according to the closeness of the state to PRIDE headquarters in Atlanta. Georgia schools were the most likely to join. PRIDE analysts applied a weighting scheme to the data to try to make their estimates nationally representative, but it could not eliminate the likely bias inherent in this highly skewed, nonrandom sample.

Another challenge to the NHSDA estimates occurred in February 1990. This time it came from another federal agency. Dr. Eric Wish, a visiting fellow at the National Institute of Justice (NIJ), had written a paper commenting on US drug policy, claiming that a large portion of criminals were heavy cocaine users, and the NHSDA was undercounting them.[5] The analysis was based on data from the NIJ's Drug Use Forecasting (DUF) program. DUF collected urine samples (which were tested for drugs) and self-report data from arrestees at participating booking facilities in twenty large cities. Wish projected that there were as many weekly cocaine users among arrestees in the sixty-one largest US cities (between 978,000 and 1.3 million) as the NHSDA had estimated were in the entire household population (862,000). Wish admitted his estimation method was crude. He pointed out that the DUF sample was not a random sample of arrestees in the nation, and that some unknown number of arrestees are included in the NHSDA estimate. Wish characterized his study as a first attempt, and invited other researchers to use more sophisticated methods to estimate cocaine use. The thrust of the Wish paper was that drug war policies should focus more on the criminal population. The paper garnered a lot of attention. Senator Joseph Biden, Chairman of the Senate Judiciary Committee, commented that "the NIDA surveys present the most optimistic picture you can find out there-and I think they especially grossly underestimate the number of hard-core cocaine users." Professor Mark Kleiman, a noted Harvard drug policy expert, said of the NHSDA self-report data "A lot of people just won't tell the nice man from the government that they smoked crack recently."[6]

Congress, with Democrats in the majority in both houses, reacted to the Wish study. The House Subcommittee on Criminal Justice, of the Judiciary Committee, called a hearing to learn more about the data discrepancies. The hearing was held on March 27, with Representative Charles Schumer presiding. The room was cramped and crowded. Testifying first was NIDA Director Schuster, followed by Eric Wish and former NIDA Director Robert DuPont. I sat directly behind Schuster, ready to help him in case detailed questions about NHSDA methodology were asked. Schuster gave a slide presentation of the latest NIDA surveys, summarizing what we knew about drug use and trends, attempting to put the DUF cocaine data in the context of the larger picture of overall drug use. Schumer thanked him "for that informative and outstanding testimony." Then Wish described the DUF program and explained his method for estimating heavy cocaine use. Schumer thanked Wish, saying "Those are incredible numbers." DuPont opened his statement with an apology, saying he had to see a patient at five o'clock and would have to leave at four o'clock. He recommended that the federal government establish a zero-tolerance policy with respect to drug use by anyone under supervision of any criminal justice agency, and that universal drug testing among that population was essential. In the question and answer period following the testimony, Schumer asked questions about NHSDA, and Schuster handled them all. Schumer asked about coverage of arrestees in the sample. In response, Schuster described the upcoming DC-MADS study, which would include criminal populations and possibly urine and hair testing. Schumer said he had received complaints from researchers that NIDA doesn't make NHSDA data tapes available for further research, and asked Schuster about this. Schuster said NIDA has a policy wherein the tapes are made available one year after they are completed. Then Schumer asked when the next household survey (the 1990 survey) will be completed, and expressed concern that the scheduling of the release would be politically motivated, for example, releasing good news just before the November election. Schuster answered "No ... Next December." Schuster went on to say results will be released every December, the sample size will be increased, and there will be a six-city oversample.[7]

Two days after the hearing, Representative Charles Rangel, Chairman of the House Narcotics Committee, issued a two-page statement about the Wish study and the administration's erroneous use of data and false claims of success. In his highly politicized statement, Rangel pointed out that according to the *Washington Post*, an ONDCP spokesman said "There is no reliable government estimate of frequent cocaine use." Rangel asserted, "If there is no reliable estimate, why is the administration saying we are turning the corner in the drug war? ... If the NIDA numbers are not reliable, why do Secretary Bennett and HHS Secretary Sullivan hold press conferences to announce the 'good news' in these surveys?"[8]

Senator Biden decided to build on the Wish study and produce a comprehensive count of hard-core cocaine addicts. The study was led by Mark Kleiman. Under his guidance, the staff of the Senate Judiciary Committee consulted with experts in the field, compiled data from a variety of different data sources (including NHSDA and DUF), made some assumptions, and combined the data. The result was an estimate of

2.2 million hard-core cocaine addicts in the United States. The report, titled "Hard-Core Cocaine Addicts: Measuring–and Fighting–the Epidemic," included an estimate for each state. The methodology was crude, relying on questionable assumptions. Indicative of the methods used, Kleiman did this work under his company's name, BOTEC Analysis Corporation. The acronym stands for "back of the envelope calculation." The report was published on May 10, along with a press release, and was widely reported in the media. Biden called for new spending for drug treatment, saying the data showed the cocaine problem is "far worse than previous guesses."[9] A few weeks after the report was published, Dr. Avram Goldstein, a leading expert in pharmacology and addiction, wrote the following in a letter to NIDA and ONDCP:

> I have just read the draft report dated May 10, 1990 prepared under the supervision of Kleiman for the Biden Committee. I am appalled at the tone and quality of this document, which is so political in its content and intention, and so dubious scientifically.

Final Design for the 1990 Survey

As Assistant Secretary Mason had ordered, the 1990 survey was conducted in the spring so that results would be available before the end of the year. Data collection began in March and ended in June, with 9,259 completed interviews, including 1,931 in the Washington, DC metropolitan area. The 100 primary sampling units (PSUs) that were used in the 1988 NHSDA sample were also used in 1990. At the second stage of sampling, additional area segments were selected within the Washington, DC metropolitan area. The response rate was 82 percent. The questionnaire was similar to the 1988 questionnaire, with some minor rewordings for clarity, a few additions, and a few deletions. Although the lead letter sent to households still did not reveal the name or topic of the survey, a new tool for interviewers to explain the survey to respondents was employed. A sheet titled "Questions and Answers about the 1990 National Household Survey on Drug Abuse" could be handed to respondents as needed.

Release of 1990 Results by President Bush

As promised by NIDA Director Schuster at the February Congressional hearing, the 1990 NHSDA results were released in 1990, on December 19. However, the bickering between the two political parties over the numbers did not diminish. In fact the rhetoric was even more intense, because it was the president who announced the findings. The event was held at 10:00 AM in Room 450 of the Old Executive Office Building adjacent to the White House. Security was tight. The small auditorium was filled, and as I entered I spotted several of the White House correspondents from the major news organizations. I found a seat near the front, next to Helen Thomas, UPI's White House Bureau Chief. President Bush came in a few minutes late, gave a short speech, and then left without taking questions. Here is a portion of Bush's remarks:

> Thank you all very much. I am delighted to be here this morning with Lou Sullivan and John Walters to announce some very encouraging news about the state of the Nation's drug problem.
>
> As you know, our administration remains fully committed to fighting this problem and stopping this scourge. And that was the promise I made to the American people in my Inaugural Address, and it is a promise that I intend to keep. And I continue to believe that the problem of drugs can be overcome with this clear national strategy and the hard work and combined efforts of millions of Americans. I am pleased to say that the news we have today suggests that our hard work is paying off and that our national strategy is having an effect.
>
> In a moment, Dr. Sullivan, my Secretary at HHS, will describe for you the results of recent surveys conducted by his Department. But I wanted to emphasize how important I believe this new information is.
>
> The national household survey and the emergency room data are the latest and most compelling evidence that drug use in America is declining significantly. And more importantly, it is declining all across the board. Overall drug use is down. Monthly cocaine use is down. Hospitals are reporting fewer drug-related emergencies. Even addictive drug use, which was once spiraling upward, has started to decline. Virtually every piece of information we have tells us that drug use trends are headed in the right direction: down. And most importantly, we are seeing these declines among the Nation's teenagers, evidence that they are learning to say no, learning to live a life free of drugs.[10]

At the end of his remarks, Bush turned the podium over to Sullivan to conduct the bulk of the press conference. Notably absent from the event was William Bennett, Bush's first drug czar. John Walters was now the acting ONDCP director. Sullivan and Walters discussed the results in more detail. A key finding that both Walters and Sullivan mentioned was a decline in weekly cocaine use, from 862,000 in 1988 to 662,000 in 1990. Unfortunately, this was not a statistically significant decline. With only about 9,000 respondents representing 200 million Americans in each of the surveys, a decline of 200,000 (one-tenth of one percent of the US population) in the estimates could easily result from just several fewer respondents reporting weekly cocaine use in 1990 than in 1988.[11] We had reviewed drafts of the press release materials and had asked ONDCP not to report it as a decline, but our recommendation had been rejected. When asked about this later, Bruce Carnes said that ONDCP does not rely solely on statistically significant data to make policy decisions.[12]

After the speeches and question and answer period, the press conference ended, and I was approached by Michael Isikoff, the *Washington Post* reporter for drug abuse. He wanted to interview me for his story on the NHSDA data. We agreed to talk on the phone later that day, after I returned to my office.

The annual DEPR holiday party happened to be on the day of the press conference. To me, it was not a problem. During the party I would take any and all calls from reporters about the survey. Division director Edgar Adams arranged for calls from reporters to be answered by NIDA's press office during the party. The party setting was in our office area, so the division secretary's master phone was manipulated to block any calls from outside during the party. But I had agreed to talk with Isikoff, so I closed my office door and connected with him. Unfortunately, the temporary phone setup caused calls on my phone to disconnect every few minutes. It was a difficult, frustrating interview. Much of the discussion focused on the statistical significance of the difference in cocaine estimates. With all of the stops and starts due to the phone problem, by the end of the interview I was not sure all of the facts were clearly communicated and understood. The next morning, Isikoff's article in the *Post*, which was also picked up by many newspapers around the country, said "Critics said the survey woefully undercounted hard-core crack abusers in inner cities, and NIDA's chief statistician conceded that the reported declines among cocaine users–described by administration officials as the most astonishing finding–were not statistically significant."[13] I was on the front page of the *Washington Post*, disputing the president, the drug czar, and the HHS secretary. Actually, Isikoff didn't get it quite right, but the misquote was understandable given the circumstances of the interview. What I had said to him was the decline in current cocaine use, from 2.9 million to 1.6 million, *was* significant. The decline in weekly use, from 862,000 to 662,000, was *not* significant. I found out that day that ONDCP had called the *Post* and requested that a clarification be printed. It was in the paper the next morning.

My concerns about statistical significance were barely noticed in the midst of the larger criticisms by more influential people. Senator Biden said the survey was "wildly off the mark" and released his own study the same day as the NHSDA release. Biden's report, an update of the earlier Kleiman study, estimated 2.4 million hard-core cocaine users. Comparing this new estimate with his May estimate of 2.2 million, Biden's report claimed there was an *increase* of 200,000 hard-core cocaine users, and contrasted this finding with the NHSDA *decline* of 200,000 weekly cocaine users. It was an interesting coincidence, especially given the tight controls within the administration on the sharing of any results before the release day. I believe there may have been a leak of the NHSDA results that influenced Kleiman's study.

Biden said the NHSDA misses more addicts than it counts, and that ex-drug czar Bill Bennett never used the NIDA data because they are inaccurate. When asked to respond to that claim, Bennett disagreed. He said that although the survey misses certain populations, it is consistent over time and the trends are correct. The trends were encouraging and the strategy was working. John Walters also defended the survey, saying "It was never intended to be a census of people who use cocaine frequently."

The 1991 NHSDA

As directed by the Assistant Secretary for Health James Mason, the 1991 survey was conducted in the spring.[14] To have results available in December, we determined that data collection had to be completed by July 1. However, because of the substantial increase in the sample size, it was cost-efficient to lengthen the data collection period. Therefore, the data collection period was January through June. The longer field period made it possible to limit the number of new interviewers needed to complete the required number of interviews. In 1990, 300 interviewers completed the survey in four months. In 1991, with the sample more than three times as large and the same field period, approximately 900 interviewers would be needed. By extending the field period to six months, the data collection was accomplished by fewer than 650 interviewers. This still required a massive field staff buildup. The final sample size was 32,594, including about 15,000 in the six oversampled metropolitan areas (Chicago, Denver, Washington, DC, Los Angeles, New York, and Miami). The interview response rate was 84 percent.

With the expansion of the sample size and annual data collection, the survey team felt it would be appropriate to update the survey target population to be consistent with other large federal surveys. So beginning with the 1991 NHSDA, the sample covered the civilian noninstitutionalized population in all fifty states and the District of Columbia. This meant the sample would cover Alaska and Hawaii. It would also be expanded to include group homes, homeless shelters, college dormitories, and other non-institutional group quarters. Definitions for these categories of residences would be consistent with those used by the Census Bureau. Residents of long-term hospitals, prisons, and nursing homes would continue to be excluded. Also, civilians living on military bases would be covered. Sampling this new target population would require new procedures and guidelines for contacting and screening potential respondents.

The questionnaire included new questions on health insurance and income, for both the respondent and the family, as requested by ONDCP. An option for proxy reporting of these items was incorporated, allowing parents to report family income and insurance information in youth interviews. Questions pertaining to perceived drug availability, drug testing, steroids, and illegal activities were added. A new module asking about receipt of treatment for drug abuse was added, with specific exclusion for cigarette or alcohol treatment. The question and answer document developed for the 1990 survey was enhanced and produced as a smaller, foldout brochure.

The 1991 NHSDA results were released at a press conference held at HHS headquarters on December 19, 1991. The findings were not as positive as in the previous year. The claims of success were more muted, and there was less political rhetoric. HHS Secretary Louis Sullivan and ONDCP Director Bob Martinez spoke. The results from the six over-sampled metropolitan areas were not mentioned by the speakers, but were in the report. The rate of current marijuana use was highest in Denver (6.3 percent) and Los Angeles (6.2 percent), and lowest in Miami (3.5 percent). There were few changes in rates between 1990 and 1991. Declines since 1985 were highlighted. Most differences between 1990 and 1991 were not statistically significant, but Martinez reported them anyway, despite NIDA's warnings. NIDA had reviewed a draft of his statement and identified numerous misstatements, such as differences in drug use rates that were not statistically significant and interpretations that could not be supported by the data. Martinez described increases in drug use among persons age 35 and older, explaining this trend indicates a cohort effect – long-term users getting older, moving into the 35-plus category. "They may also very well be former drug users who are relapsing," he speculated. A chart presented at the press conference showed the number of weekly cocaine users had increased from 662,000 in 1990 to 855,000 in 1991 (it was not statistically significant), but only among the 35-and-older age group.

This would be the final NHSDA release by the George H. W. Bush administration and by NIDA as well. A couple weeks after the press conference, the new sampling plan with continuous data collection would begin. Interviewing would be taking place every month for the next quarter century and beyond. With a twelve-month data collection period in 1992, results would not be available until mid-1993, under a new administration and a new agency. However, there was still plenty of time for the Bush administration and their opponents to squabble over drug abuse statistics.

Competing Drug Control Strategies

Soon after the release of the 1991 results, the political attacks resumed. ONDCP released the 1992 National Drug Control Strategy on January 27, 1992. On the same day, Senate Judiciary Committee Chairman Joe Biden sent out a press release criticizing the strategy:

> Today's strategy contains the same flaws–the same failure to make the tough decisions–that have brought us to the current stalemate in the "war on drugs." Unless there are some changes in direction, our nation's drug epidemic is going to continue to grow in the years ahead.

Biden's press release announced he would be producing his own drug control strategy, with his recommendations to bring the drug epidemic under control. His "hard-core cocaine addict" estimates were cited again. But there was a new concern. He referred to the administration's own data showing a rising heroin problem: 1.2 million Americans started to use heroin in 1991, a 75 percent increase over 1990. Biden's announcement puzzled ONDCP, and they immediately asked ADAMHA to explain where these numbers came from and if they were accurate. The ONDCP inquiry made its way through the bureaucracy (at the time, ONDCP was not permitted to send questions or data requests directly to me), finally reaching me at about 3:00 PM on January 28. A response was needed by the end of the day. Fortunately, we had just completed a research paper on methods for estimating drug use incidence, or first time use.[15] We had the answer, but needed to explain it clearly, in writing, so that it could be transmitted back through the appropriate reviewers and approvers in NIDA, ADAMHA, HHS, and finally to ONDCP. The estimate that Biden cited was a simple calculation using published NHSDA numbers. The estimates of the number of persons who had used heroin in their lifetime were 1,654,000 in 1990 and 2,886,000 in 1991. The difference between these two estimates is 1.2 million, an estimate of the number of people who used for the first time between 1990 and 1991. The problem with this calculation is that it is subject to extreme variability, due to the small sample sizes.[16] In fact, the 95 percent confidence interval for Biden's estimated number of new users was about 500,000–2,100,000. But even the lower bound of this confidence interval was more than five times the annual estimates for the 1980s based on the "retrospective" estimation method.[17] If Biden's calculation had been used to estimate the number of new users between 1988 and 1990, the result would have been an impossible *negative* 250,000. As for the "75 percent increase over 1990," that number reflects the increase in the number of lifetime users, not an increase in the number of new users, as incorrectly stated in the press release.

Controversy over Revision of a Key Cocaine Indicator

In early February 1992, while running some exploratory analysis on the 1991 NHSDA data file, I made a troubling discovery. I found eleven cases on the file that were age 40 and older and coded as weekly cocaine users as a result of the hot-deck imputation applied to this variable (see discussion in Chapter 3). This concerned me because in 1990 there had been no such cases, and because these eleven respondents to the 1991 survey had a combined weight of 210,000, which means they represented about two hundred thousand adults – almost exactly the size of the "increase" in weekly cocaine use among older adults cited at the December press conference, and one-quarter of the estimated number of weekly cocaine users. Was it possible the contractor had changed the imputation procedure in 1991, resulting in an estimated 210,000 additional weekly cocaine users, all age 40 or older? The large impact of imputed data on the overall estimate and the trend is a red flag indicating too much missing data or a poor imputation method may be resulting in severely biased estimates. I immediately contacted RTI and asked them to investigate further. RTI reported back to me that yes, there had been a change in procedures. It affected the frequency of use variables for ten drugs. Although the same basic hot-deck procedure that had been used in 1990 was repeated in 1991, they had added a new sort variable because of the six-city oversample. RTI had made this change to ensure that an imputed value would come from a donor in the same geographic area. For example, if a respondent in Denver did not report their frequency of use, the missing information for that respondent would be replaced in the data file by the frequency of use information from another Denver respondent. The idea of keeping imputation geographically controlled was reasonable, but there was no follow-up check on how well this modification worked, or whether it affected the national estimates differently in 1991 than in prior years, until it was too late.

By the end of February, RTI had revised the imputation, created new frequency of use variables on the data file, and generated replacement tables for the Population Estimates report. The report had previously been printed and distributed. Shown in Figure 5.1, the revised weekly

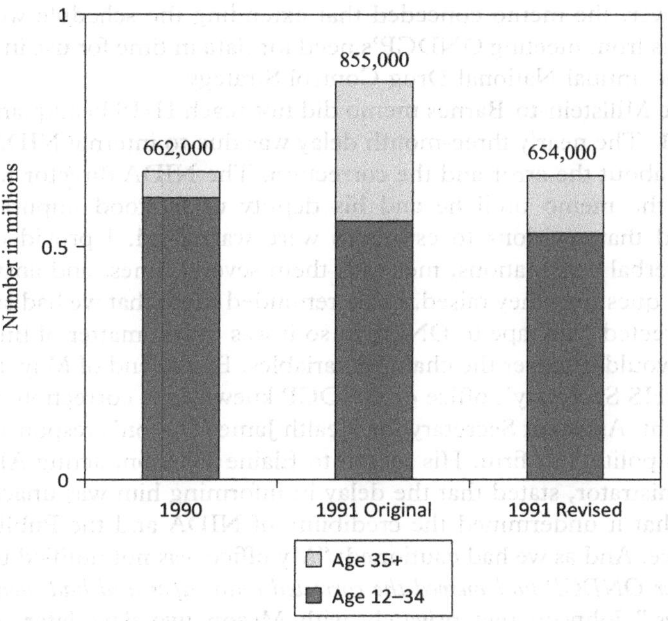

Figure 5.1 Weekly cocaine use by age: 1990 and 1991, before and after the imputation revision.

cocaine estimate was 654,000, a 24 percent reduction from the previously published number.

Estimates for most other drugs were not affected much. In early March, I drafted a memo to inform HHS and ONDCP of the revisions to the estimates. The memo was from Richard Millstein, Acting NIDA Director, to Mark Barnes, HHS Counsel to the Secretary for Drug Abuse Policy. And of course it would have to travel through the bureaucracy and be approved by the necessary senior officials before reaching Barnes. The detailed, four-page memo explained how the error occurred and showed the revised estimates. It also reminded Barnes that ONDCP had highlighted the weekly cocaine use estimate at the December press conference, claiming an increase from 1990 to 1991 despite the lack of statistical significance. The memo described new data processing quality control procedures we would employ in future surveys, and stated that the best course of action to reduce the risk of errors was to lengthen the time for the contractor and the small project management team in DEPR to do a thorough check of the data before release. Five months between the end of data collection and the release of the data was just too short.

However, the memo conceded that extending the schedule would prevent us from meeting ONDCP's need for data in time for use in developing the annual National Drug Control Strategy.

The Millstein-to-Barnes memo did not reach HHS headquarters until June 1. The nearly three-month delay was due to internal NIDA discussions about the error and the correction. The NIDA director would not send the memo until he and his deputy understood imputation and agreed that revisions to estimates were warranted. I provided written and verbal explanations, met with them several times, and answered all of the questions they raised. I also reminded them that we had submitted a corrected data tape to ONDCP, so it was only a matter of time before they would discover the changed variables. By the end of May, no one at the HHS Secretary's office or ONDCP knew of the correction. Or so we thought. Assistant Secretary for Health James Mason's response on June 9 was polite, but firm. His memo to Elaine Johnson, acting ADAMHA administrator, stated that the delay in informing him was unacceptable, and that it undermined the credibility of NIDA and the Public Health Service. And as we had cautioned: "My office was not notified until June 1–*after ONDCP had viewed the corrected data tapes and had alerted Mark Barnes*." Johnson met privately with Mason two days later, and then followed up with a three-page apologetic memo back to Mason on June 12, saying "In retrospect, this took longer than would be ideal, but it was NIDA's intention to ensure that the revised procedures were justified and valid."

Not long after the corrected frequency data were distributed, another problem was discovered in the 1991 data. NIDA's timing in informing the White House was even worse this time.

Pre-Election Data Correction

After the discovery of the frequency-of-use imputation problem in February 1992, evaluation of the entire 1991 NHSDA data processing and estimation continued for several months. Another unusual pattern emerged. This time it was heroin. An unusually large number of cases were coded as heroin users based on the logical editing. At first, this did not alarm us because in contrast to imputation, editing rules could only determine that a respondent used a drug if they actually reported use somewhere in the questionnaire. But we eventually realized that it was not always safe to rely on that rule.

When NHSDA editing was automated in 1988, the programs were set up to be consistent with hand-editing guidelines used by RAC for previous surveys. The rule in determining if a respondent had used a drug was that any evidence of use is sufficient to be labelled a drug user. For example, five denials of use in one section of the questionnaire but one report of use in another section would result in coding as a user. DEPR and RTI analysts were not comfortable with this editing principle, and we were planning on switching to a new rule in the 1993 or 1994 survey. But for 1991 data, as with 1990 and all prior years, the rule was "any evidence."

In the 1990 survey, a question in the alcohol module asked which drugs were used at the same time as alcohol during the past twelve months. The question (number A-11) appeared at the top of a page in the alcohol use module. A list of nine drug categories appeared beneath the question, and to the right of each drug name was a number for respondents to circle if they had used that drug with alcohol. If a respondent indicated use of a drug here, then editing would classify that respondent as a past year user of the drug, regardless of whether or not they had reported use of the drug in another section. In the 1991 survey, the same question was asked, and the same editing rules were applied. But in 1991, the question was moved from the alcohol section and placed instead in a "Drinking Experiences" module – immediately below a series of questions about experiences with drinking in the past twelve months in which the response options provided to respondents to check were "Yes" or "No." Shown in Figure 5.2, "No" responses were on the far right of the page. Directly below these "No" responses were the "Yes" response options to the alcohol-drug combination question. The layout apparently caused some inattentive respondents who had not used illicit drugs to mistakenly circle the "Yes" answers for all nine illicit drug categories.

This was a classic case of poor questionnaire construction. Question DE-2 in the 1991 survey was exactly the same as question A-11 in the 1990 survey. But moving it to a different place in the questionnaire affected how respondents answered. In combination with the "any evidence" rule, it apparently led to fifty-three respondents being incorrectly coded as past year heroin users. When the editing was corrected and revised estimates were revealed, the number of past year heroin users was changed from 701,000 to 381,000. Inspection of the fifty-three problematic cases revealed that one of these respondents was a 79-year-old woman with an analysis weight of 142,000. She represented 142,000 79-year-old women. The lifetime heroin use estimate (which Senator Biden had used in his incidence calculation in January) was lowered from 2,886,000 to 2,653,000. Estimates for other drugs were also corrected, although the relative effect was not as large because the prevalence rates were much higher for the other drugs.

<u>In the past 12 months</u>, ...		YES	NO
m.	I have awakened unable to remember some of the things I had done while drinking the day before	01	02
n.	I had a quick drink or so when no one was looking	01	02
o.	I often took a drink the first thing when I got up in the morning	01	02
p.	My hands shook a lot after drinking the day before	01	02
q.	Sometimes I got high or a little drunk when drinking by myself	01	02
r.	Sometimes I kept on drinking after promising myself not to	01	02

DE-2. **In the <u>past 12 months</u>, what drugs listed below did you use on your own, that is, nonmedically, at the same time or within a couple hours of when you drank beer, wine, or liquor? (PLEASE CIRCLE ALL THAT APPLY.)**

Sedatives (barbiturates, sleeping pills, Seconal ("downers"))	01
Tranquilizers (antianxiety drugs like Librium and Valium)	02
Stimulants (amphetamines, Preludin ("uppers" or "speed"))	03
Analgesics (pain killers like Darvon, Demerol, Percodan, Tylenol with codeine)	04
Marijuana	05
Inhalants (glue, amyl nitrite, "poppers," aerosol sprays)	06
Cocaine (including "crack")	07
Hallucinogens like LSD, PCP, peyote, mescaline	08
Heroin	09
Used alcohol but did <u>not</u> use any of these kinds of drugs in the past 12 months at the same time or within a couple hours of drinking alcohol	10
Used alcohol in your life but did <u>not</u> drink beer, wine, or liquor in the past 12 months	93
Never had a drink of beer, wine, or liquor in your life	91

(PLEASE TELL THE INTERVIEWER WHEN YOU ARE FINISHED)

2

Figure 5.2 Page 2 of Drinking Experiences answer sheet, 1991 NHSDA.

In the months just prior to the 1992 presidential election, drug abuse policy was a prominent topic in campaign rhetoric. Democrats Biden, Rangel, and Schumer criticized the administration's handling of the drug war. Vice-presidential candidate Al Gore, campaigning in

Louisiana, joined in. He accused Republican incumbent candidate Bush of appointing a "defeated political crony" to run ONDCP.[18] Gore's campaign aides pointed out to reporters that September 5 marked the third anniversary of Bush's nationally televised speech in which he held up a bag of crack from a staged purchase across the street from the White House. Drug czar Bob Martinez was heavily involved in campaigning for Bush, attacking Democratic presidential candidate Bill Clinton's record on drugs as weak because Arkansas (where Clinton had been governor) allowed convicted drug sellers and users to get driver's licenses.

On September 10, the Senate Judiciary Committee, chaired by Senator Biden, held a hearing to announce an update of their drug abuse policy report. The report was critical of the administration's policies and cited increasing heroin use as evidence of its failure. Drug czar Martinez testified at the hearing. Unfortunately, NIDA's memo informing HHS of the revised heroin estimates did not reach ONDCP until September 11, *the day after* Biden's hearing.

The memo downplayed the impact of the revision as minor. Even the revision of the 1991 heroin estimates for 1991 was said to not be a concern, because the change from 1990 was not statistically significant before or after the correction and the precision for heroin use estimates "is not particularly good." There was no apology and no mention of the Judiciary Committee hearing.

A few days later, HHS Secretary Sullivan received an unhappy memo from Martinez. Here are some excerpts:

> The Household Survey on Drug Abuse, for better or worse, is the report card by which the Administration's progress in the drug war is measured. The 1991 Survey contains estimates of major increases in cocaine and heroin use that, when released, were used to attack the Administration mercilessly. At our hearing before the Senate Judiciary Committee on September 10th, for example, Senator Biden cited the published figures on increases in the number of lifetime heroin users to argue that the Administration is losing the drug war, a claim made by the media and the Clinton campaign.
>
> We appreciate the Survey's natural limitations with respect to scope and coverage, but we cannot understand why it is plagued with problems in the data editing and imputation, and why it takes so long to verify and correct the data.
>
> Further, we cannot take refuge in the issue of statistical significance to argue that the revised data are essentially the same as the published data. These estimates are used by the media and the Congress to draw inferences about drug trends regardless of the issue of statistical significance. Had we been given the revised estimates on September 10th, we could have better defended the Administration against Senator Biden's attack.

Sullivan sent a reply to Martinez, but not until November 30. The election was over, Bill Clinton would be president in a few weeks, and both Sullivan and Martinez would soon be out of office. Sullivan's memo was apologetic, but also blamed the episode on ONDCP's insistence on receiving data early, before it was fully checked. "We cautioned ONDCP at that time that the data were preliminary and subject to change." Sullivan also listed some steps that would be taken to prevent future problems:

- An editing and imputation quality control report produced each year prior to finalization of estimates.[19]
- New imputation methods for the 1992 data.
- A new questionnaire, under development, would reduce missing data and inconsistencies.
- Improved editing procedures for the new questionnaire.[20]

And with that, the turbulent Bush-Bennett-Martinez-Carnes-Sullivan-Mason-Barnes era came to an end. Despite the conflict and miscommunication, the long-term impact of this era on the National Survey was positive. The larger sample size would strengthen the survey's ability to track trends accurately. It would also expand the capability of the survey to support in-depth studies of special populations and rare behaviors. Ironically, the ONDCP officials that regularly downplayed the importance of statistical significance in reporting the survey results were also responsible for increasing the sample size to improve statistical precision. ONDCP identified weaknesses in the survey and pushed NIDA to make improvements in the methods. They supported methodological research, despite their limited understanding of how differences in methodology can affect estimates. With their concurrence, NIDA was able to conduct a major series of research studies on NHSDA methods, focused on improving the survey data. Some of this work is described in the next section.

NHSDA Methodological Research

The program of methodological research on NHSDA began in 1989, when $1.6 million was added to the 1990 survey contract for this purpose. The program focused on three main areas: questionnaire administration, nonresponse, and sample design. Preliminary results were presented to a panel of outside experts and ONDCP at a "Methodology Workshop" in April 1991 (see Chapter 4). Most of this initial set of studies was completed by the end of 1991. The results provided a basis for a new questionnaire that was eventually fielded in 1994.

The 1994 survey also included a simultaneous supplemental sample that administered the "old" questionnaire, providing a way to measure the impact of the changes and possibly preserve trends.

In-depth evaluation of the questionnaire was done with cognitive studies and analysis of responses to prior surveys. The cognitive studies involved systematic assessment of survey items, identifying vague or ambiguous terms, unclear questions, undefined reference periods, confusing formats, and questions creating a difficult cognitive task, such as remembering events from a long time ago or involving complex definitions. "Think aloud" sessions asked respondents to explain their thought processes as they answered NHSDA questions. Laboratory experiments were conducted to test alternative questioning strategies. Analysis of missing, invalid, and inconsistent responses helped identify questions that were difficult for respondents to answer. This analysis also provided useful information on the impact of editing and imputation, and helped in developing improvements to these processes. Based on the 1988 data, it showed the huge impact that using the alcohol-in-combination data in editing had on heroin estimates, even before the combination question was switched from the core alcohol module to the "Drinking Experiences" module in 1991. Results from all of these questionnaire studies fed into the design of a new, improved questionnaire, which was tested in a 4,326-respondent field test experiment, conducted in the fall of 1990. Half of the sample was given the old (1990 NHSDA) questionnaire and half was given the new one. Further, each of the half-samples was split in half to test another aspect of the questionnaire administration, the mode: self-administered answer sheets versus interviewer-administered questioning, for the drug use questions. In experimental design terms, it was a 2 x 2 factorial experiment. The most important result from the field test was the mode effect on reporting of drug use. Higher rates of drug use were obtained with self-administered questions than with interviewer-administered, and the difference was greatest for the most recent time period and for the most sensitive or socially undesirable drugs. Lifetime alcohol use rates were equal in the two modes, but for past thirty-day cocaine use, the estimated rate of use was 2.4 times higher with self-administered than with interviewer-administered. A limited evaluation of skip patterns was also included in this 1990 field test. In one self-administered questionnaire version, respondents were instructed to skip over a set of questions about use of a drug if they answered initially that they had not used that drug. Nearly all respondents correctly followed these instructions. The 1990 field test was not designed to test whether rates of use were affected by having skips within the questionnaire, but in 1992 an experiment found that skip patterns resulted in lower (and presumably less accurate) prevalence estimates for marijuana and cocaine use.[21]

Person and household-level nonresponse patterns and bias[22] were analyzed in several studies. A detailed analysis of nonresponse patterns in the 1988 NHSDA showed which types of respondents were most likely to refuse to participate or not be available to participate in the survey when asked. There was no consistent pattern of high drug using population subgroups being more reluctant to participate. In fact, some of the groups with the lowest rates of drug use (for example, elderly persons) had relatively low response rates. So overall, there was no clear conclusion of the effects of nonresponse on the accuracy of NHSDA estimates of the prevalence of drug use. The overall direction or magnitude of nonresponse bias could not be definitively determined without more data on the nonrespondents. A small nonrespondent follow-up study was done in 1990 to learn more about nonrespondents. In this study, a sample of nonrespondents in the Washington, DC area was tracked down after the survey was completed, and asked again if they would be willing to answer a few brief questions about drug use. To convince these initial nonrespondents to participate, they were told it would only take a few minutes because the questionnaire was greatly reduced, and they would be given ten dollars. A response rate of only 38 percent was achieved and the sample size was only 144, so the findings were not definitive. No statistically significant differences in rates of past month or past year drug use were found between the initial survey respondents and initial non-respondents who agreed to the follow-up interview.

Another study of nonresponse was the Census Match study. Conceived and coordinated by Dr. Robert Groves and Dr. Mick Couper, this study involved linking nonresponding cases from NHSDA and several other large federal surveys with their Decennial Census data.[23] This gave us demographic and other data (not substance use) for over 90 percent of nonrespondents in the 1990 NHSDA. Similar to the findings of the nonresponse analysis, the Census Match study found that some populations with low response rates tend to have high rates of drug use (male, large metro area), while other low-response rate populations tend to have low rates of drug use (persons older than 35, high income, white).[24]

Research on the sample design was done through the RTI contract under a "Design Issues" study. This study was actually a series of studies that provided some basic data on the efficiency of the current design, and explored the potential effects on cost, precision, and data quality of modifications to the sample. Ideas for increasing the number of drug users in the sample were assessed. One approach that emerged and was later tried in the 1993–1995 surveys was adding cigarette use questions to the household screener and assigning higher selection probabilities to household members reported to be current smokers.

We recognized the importance of sharing the results of the methodological studies with the survey research community. Although our studies were specifically tailored to NHSDA concerns, many of the findings would be applicable to other surveys. Also, since the studies played a role in the 1994 redesign, and could potentially help in making future design decisions, it was appropriate to make reports on them available. The results would be important background information for federal and contractor staff working on NHSDA in the future. Findings about specific data items would help researchers working with the NHSDA data files in determining whether and how to use those data items for their studies. Dr. Charles Turner at RTI suggested that we compile the results and publish them as a book. NIDA approved, and decided it should be a NIDA publication. Several thousand copies were printed, and NIDA distributed free copies to many statisticians and drug abuse researchers.[25]

We also organized a session at the 1992 American Statistical Association Conference, in Boston. The seventy-five-minute session included presentations by NIDA and RTI analysts on the NHSDA methods research. It was the first time the survey had a significant presence at a major statistical meeting. The benefits of attending and presenting at conferences were clear, and the survey team, including contractor staff, participated in many conferences like this in subsequent years. Methodological research became a regular component of the NHSDA project. Every large ongoing survey needs this. I made sure each drug survey contract included tasks such as "ad hoc methods studies" and "field tests" with adequate staff and funding. After we spent the $1.6 million on methods studies in the 1990 contract, there were still more studies needed. The NHSDA team in NIDA had ideas, as did RTI and ONDCP. There was also an ongoing need to be able to quickly investigate unusual and unexpected findings and to check for possible errors in the data. Potential methodological studies included a validation study obtaining hair and urine from respondents to test for drugs and compare with self-report; a field test of the effects of giving survey respondents a monetary incentive for participation; a diagnosis study that would administer the NHSDA questionnaire to a sample of clients from drug abuse treatment facilities to improve the capability of the survey to identify persons who need treatment; a reliability study, in which NHSDA respondents were re-interviewed a week or two later to measure how consistently they responded to each survey item; and a nationally representative nonrespondent follow-up study. Eventually, all of these studies would be implemented except the last one.

NIDA and Its Grantees Continue to Oppose the Large Sample

Late in 1991, proposed legislation in Congress called for reductions in NIDA's budget. NIDA's National Advisory Council sent a letter to NIDA Director Schuster recommending that to accommodate the budget cut, NIDA should begin immediate steps to (1) reduce the size of the survey, (2) delete from authorizing legislation the requirement for an annual household survey, and (3) seek innovative ways to fund the metropolitan area studies and reduce their cost. The council said "The marginal scientific utility of such a large sample in terms of precision of estimates is small." Then NIDA convened a group of outside experts to obtain input.[26] Their recommendations were varied. James Anthony, Lloyd Johnston, and Richard Clayton agreed that a sample of 8,000 to 10,000 every other year would be sufficient to track trends. Johnston also suggested that NIDA consider dropping the age 12–17 sample from NHSDA, assuming his study, Monitoring the Future, would continue to cover youths. Ronald Wilson, Director of the Division of Epidemiology and Health Promotion at NCHS, said "Given the continued importance of the drug problem in this country and its link to HIV, the NHSDA should continue with the largest sample possible. Any 'war on drugs' should be driven by the best data available. The current cost of $10 million is hardly extravagant."

A memo from ADAMHA Acting Administrator Elaine Johnson to Assistant Secretary for Health James Mason outlined NIDA's final recommendation: Conduct the survey every other year in the spring, with a national sample of 12,000, and no oversampling of metropolitan areas.

No action was taken in response to the memo. Perhaps it was because of the March 4, 1992 note from ONDCP to HHS, urging that *no cuts* be made to the survey. Or it could have been because the ADAMHA Reorganization Act was nearing passage. This bill would break ADAMHA apart, placing much of NIDA, NIAAA, and NIMH in the National Institutes of Health, with the remainder of ADAMHA, including NHSDA, forming a new agency, the Substance Abuse and Mental Health Services Administration (SAMHSA).

Preparation for the 1993 survey continued, with a planned sample size of 26,000. Decisions on the budget and the amount to be spent on the survey would wait for SAMHSA leadership to be in place. SAMHSA would begin its existence on October 1, 1992, after which NHSDA would no longer be NIDA's responsibility.

Notes

1 General Accounting Office, *Drug Use Measurement.*
2 Isikoff, "Survey Finds Drug Use Underreported." *Washington Post*, October 1, 1989.
3 Since 18- and 19-year-old twelfth graders would not be counted in the NHSDA estimates, the comparison would be expected to show a greater number of users in the PRIDE data than in the NHSDA data.
4 Gfroerer, Wright, and Kopstein, "Prevalence of Youth Substance Use," 19–30; Gfroerer, "Influence of Privacy," 22–30.
5 Wish, "Drug Policy in the 1990s."
6 *US News and World Report*. March 5, 1990.
7 Transcript of the Oversight Hearing, March 27, 1990 regarding Emerging Criminal Justice Issues: Drug Use and the Pretrial Population.
8 Press release from Congressman Charles Rangel, March 29, 1990.
9 Isikoff, "Senate Study Triples Cocaine User Estimate." *Washington Post*, May 11, 1990, A4.
10 Bush: "Remarks at a White House Briefing."
11 In this case, sixty-five respondents reported weekly or more often cocaine use in 1988, and sixty-one reported it in 1990.
12 GAO, Drug Control: Reauthorization, 57–59.
13 Isikoff, "US Survey Shows Sharp Drop in Illegal Drug Use." *Washington Post*, December 20, 1990, A1.
14 James Mason, memo to Administrator, ADAMHA, October 16, 1989.
15 Gfroerer and Brodsky, "Incidence of Illicit Drug," 1345–51.
16 The sample size in the 1991 survey was 32,594, but the overall "effective" sample size was only about 5,800 for the lifetime heroin use estimate. This was due to a large design effect resulting from the oversampling of young people, blacks, Hispanics, and the six metropolitan areas, as well as the cluster sampling in the multistage complex sampling plan. Similarly, the effective sample size in 1990 was about 5,200, despite the actual sample size of 9,259.
17 The retrospective method provides more stable estimates by using data on age at first use, along with interview data and date of birth of respondents, to construct annual incidence estimates.
18 Martinez was governor of Florida, but lost his bid for reelection in 1990, and was given the drug czar job by Bush in March, 1991. After taking office, he arranged to have refunds from his gubernatorial race campaign funds rerouted to the Bush-Quayle reelection campaign.
19 This report became a contract deliverable for every survey, starting with 1992. It consisted of a set of tables of preliminary data comparing unedited, edited, and imputed estimates for the current year and prior years, highlighting any changes in the impact of editing and imputation.
20 At the time the memo was written, we had not yet determined which year the new questionnaire and editing rules would be fielded. We eventually implemented it in 1994, with a supplemental sample employing the old questionnaire, so that we could measure the impact on estimates and continue assessing trends.

21 Gfroerer, Lessler, and Parsley, "Studies of Nonresponse," 273–295.
22 Nonresponse bias for a survey estimate of drug use prevalence would occur when nonrespondents have a different rate of drug use than respondents. The magnitude of the bias depends on the response rate and the difference in prevalence rates between respondents and nonrespondents.
23 Groves and Couper, *Nonresponse in Household Interview Surveys.*
24 Gfroerer, Lessler, and Parsley, "Studies of Nonresponse," 273–295.
25 Turner, Lessler, and Gfroerer, *Survey Measurement.*
26 Consultants were James Anthony (Johns Hopkins University), Richard Clayton (University of Kentucky), Lloyd Johnston (University of Michigan), Peter Reuter (RAND Corporation), Richard Spence (New Jersey), and Ronald Wilson (NCHS).

Wrap-Up for Chapters 3, 4, and 5

During the 1980s, the survey was revised slightly each time it was conducted, due in part to reorganizations within NIDA and shifts in oversight responsibility. Survey design and content changes are sometimes influenced by the area of interest of the agency or division responsible for the survey, as well as the personal opinions of staff who directly work on the survey. After oversight of NHSDA moved from NIDA's research division to its medical affairs division, the use and effects of prescription drugs became a priority. The module on prescription drug use was expanded to cover medical use, and the effects of use. To improve data quality, the nonmedical prescription drug use questions were converted from interviewer-administered to self-administered in 1982, causing a potential trend break. Tobacco and inhalants were lower priority for the medical affairs staff, and they removed inhalant questions from the questionnaire. After residing with the medical affairs group for a short time, the survey was placed in an epidemiology division, along with other data collection and analysis activities. That division was staffed with statisticians and epidemiologists. In that environment, statistical issues such as survey design, estimation methods, statistical integrity, and coordination with other data systems received more attention. Several improvements were implemented. External expert consultants continued to play a major role in decisions about the questionnaire content and survey design during this period.

As had been the case when the survey began in 1971, external social and political influences led to the most important changes in the survey in the early 1990s. Reacting to the widespread use and consequences of crack cocaine, and the cocaine overdose deaths of Len Bias and Don Rodgers, Congress passed legislation that expanded NHSDA and other data systems, and increased research funding. The legislation also created ONDCP, stipulating that they must develop and track quantified objectives that measured progress towards reducing drug use. Soon after ONDCP was created, officials there identified NHSDA as a primary source of data for these objectives. They also recognized that the survey would have to be modified to fully satisfy this need. The sample had to be larger, and the questionnaire needed revision. The urgency and authority with which ONDCP pursued these changes created conflict between ONDCP and NIDA. A driving force behind the conflict was the objections by NIDA leadership and grantees concerning the cost and benefits of the expansion. But the survey team also had concerns about statistical integrity. ONDCP staff did not have backgrounds in statistics or survey research. We thought that some of the changes they demanded could be detrimental to the survey integrity or would not produce valid and reliable data. We also thought other changes might be workable, but should undergo some development and testing before implementation. The impatience of ONDCP staff in pushing for these changes may have been partly due to their lack of appreciation for the amount of time it takes to plan and conduct a large national survey. Or perhaps it was just their enthusiasm in their important new roles helping to direct the nation's war on drugs. I suspect it was also partly a function of the election cycle. The White House surely wanted to accumulate some accomplishments on the drug war in time for the next presidential campaign in 1992.

Conflict between ONDCP and NIDA led to even more communication difficulties. ONDCP's arrogance created some animosity between ONDCP and HHS officials at high levels, and to protect the NHSDA project and limit ONDCP's control over NIDA staff, HHS created bureaucratic roadblocks and protocols that hampered information flow. Some problems (for example, NIDA's delay in informing ONDCP about the correction to the frequent cocaine use estimate; ONDCP's misunderstanding of "non-response") could have been averted if these barriers had not existed. But these problems might also have been less severe if we had provided better explanations and translations of statistical and survey research concerns to non-statisticians at ONDCP and NIDA.

NIDA's concerns about the integrity of the survey were not limited to the design changes that ONDCP was intent on accomplishing. We were also frustrated with the ways ONDCP was reporting the results. In each of the first three NSDUH data releases they were involved in (1988, 1990, and 1991 survey data), the drug czar's speech included claims of changes in drug use rates that we had advised ONDCP not to report, because the differences were not statistically significant. With the 1988 data release, ONDCP's message was that the cocaine problem was getting much worse. Of course, what else could they say, given that the office had just been created to combat the problem? But although cocaine overdoses and treatment admissions were rising, most of the indicators from NSDUH showed declines. So drug czar William

Bennett highlighted the weekly cocaine use estimate, which was higher in 1988 than in 1985 (but not significantly) and characterized the change as "a shocking and unacceptable jump." Ironically, the extensive media coverage of the Bias and Rodgers overdose deaths, which triggered legislation that created ONDCP, was probably what scared many recreational cocaine users into quitting their use, resulting in substantial declines in overall cocaine use just before the 1988 survey. In reporting the 1990 survey results, there was more willingness by the White House to report declines in drug use. They had been in office for nearly two years, and the president could now report that "our national strategy is having an effect." This time, the survey showed a smaller number of weekly cocaine users in 1990 than in 1988, but it was again not a statistically significant difference, and again we advised ONDCP not to report it as a decrease. But acting drug czar John Walters did just that. A year later, drug czar Bob Martinez also ignored our recommendations and reported numerous interpretations not supported by the data.

Despite ONDCP's overtly political and incorrect uses of the data and reckless demands to immediately redesign the survey, there were also signs that ONDCP staff recognized the importance of statistical integrity, and the value of science-based policymaking. Through their discussions with the project staff and review of the reports and background materials we shared with them, they must have recognized the quality of the survey methodology and the transparency of the project, which may have influenced their decision to invest in the project by expanding the sample. They supported the new program of methodological studies that NIDA had begun. After considering our reasons for opposing their fast-track questionnaire changes, they did back off of demanding many of those unreasonable changes, allowing more time for us to carefully update the instrument. And even when ONDCP selectively highlighted a particular trend in the cocaine data (frequent cocaine use as a percentage of past year cocaine use) in their first interim drug control strategy, they verified with me that the change in the estimate between 1985 and 1988 was statistically significant.

6 The Survey Moves to SAMHSA

The summer of 1992 was a stressful time for the staff in NIDA's Division of Epidemiology and Prevention Research (DEPR). The much anticipated ADAMHA Reorganization Act had finally been passed on July 1, and leaders in the agency were scrambling to finalize the details, including the placement of DEPR projects and the people who managed them. The primary purpose of the reorganization was to separate research activity from services activity. Grantees, advocacy groups, professional associations, and members of Congress had voiced concerns about the ability of a single agency (ADAMHA) to oversee both research and service-related activities. They raised questions about priorities and funding levels, and expressed a need to improve the transfer of the results of basic research into clinical practice.[1]

The Creation of SAMHSA and the Office of Applied Studies

The ADAMHA Reorganization Act specified that research components of ADAMHA's three institutes (NIDA, NIAAA, and NIMH) would be placed in the National Institutes of Health, and the service-related functions in the institutes and other parts of ADAMHA would be placed in a new agency, the Substance Abuse and Mental Health Services Administration (SAMHSA). The administrator of SAMHSA would be appointed by the president, and approved by the Senate. SAMHSA would contain three centers: The Center for Substance Abuse Treatment (CSAT), the Center for Substance Abuse Prevention (CSAP), and the Center for Mental Health Services (CMHS). A major responsibility of SAMHSA would be the management of the block grants that provided funds to states each year for the prevention and treatment of substance abuse and mental illness. These block grants would account for most of SAMHSA's budget. The largest was the $1.5 billion Substance Abuse Prevention and Treatment (SAPT) Block Grant. CSAT would manage this program. A smaller ($450 million) Community Mental Health Services Block Grant was to be managed by CMHS.

During the drafting of the legislation, there was uncertainty about which agency the NHSDA and other NIDA data systems would be placed in. NIDA and DEPR leaders wanted the surveys to remain in NIDA, as part of the agency's epidemiology program. In their view, data collection was an integral component of epidemiological research programs, and similar programs in other NIH institutes had large surveys. NIDA director Schuster explained NIDA's position to NIH Director Dr. Bernadine Healy, who said she agreed. However, after meeting with Healy, Schuster told DEPR acting director Zili Amsel in an e-mail "We have to realize that ONDCP does have some statutory authority over the reorganization plan and must certify it. Therefore, even with the support of Dr. Healy it is likely that we will lose certain functions." Anticipating easier access to the data, ONDCP wanted all of ADAMHA's major drug abuse data systems to be in SAMHSA, and that was the final outcome, with one exception.[2] NIDA leaders lobbied for and were successful in retaining the Monitoring the Future (MTF) study, in part because it was funded through a grant. NHSDA, DAWN, NDATUS, and TEDS were contract activities.

The ADAMHA Reorganization Act did not specify where in the new agency the substance abuse data systems would reside. However, the bill did indicate that funding for this data collection would come from a 5 percent set-aside in the SAPT Block Grant. In other words, $75 million per year (5 percent of $1.5 billion) would be "tapped" from the Block Grant to fund national substance use data collection, as well as technical assistance to states and evaluations. There was also a 5 percent set-aside for the mental health block grant, to fund national mental health data collection. The mental health set-aside would be managed by CMHS, but the substance abuse data collection would not be managed by CSAT, the center that was responsible for implementing the substance abuse block grant. A separate Office of Applied Studies (OAS) was created for substance abuse data. OAS would be part of the SAMHSA Office of the Administrator, to formally acknowledge the importance of the surveys and ensure responsiveness to ONDCP requests for information. The placement of the major drug abuse data systems before and after the reorganization is displayed in Figure 6.1.

Unfortunately, these decisions on placement of SAMHSA's data systems led to problems with working relationships across SAMHSA components. Some CSAT managers and staff felt that since they would be responsible for managing the SAPT Block Grant, the set-aside should also be managed by CSAT, including authority and oversight

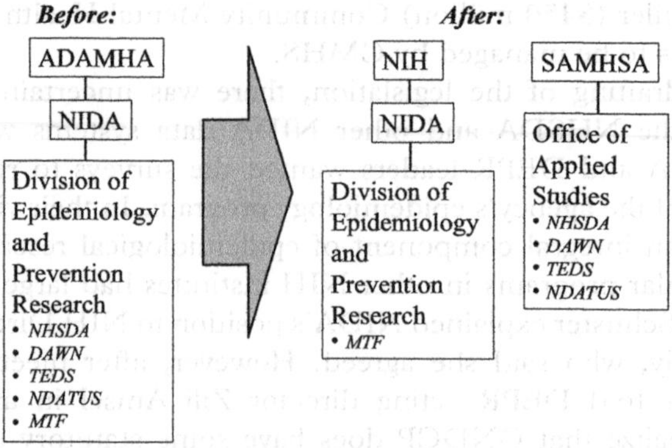

Figure 6.1 Location of drug data sets before and after creation of SAMHSA, 1992.

responsibility for NHSDA. Leaders and staff at CMHS, on the other hand, seemed to want no involvement with OAS or NHSDA. They had their set-aside funding and their own mental health surveys, and had little interest in collaborating with OAS. Unless pushed by the administrator, CMHS resisted plans to add mental health items to the NHSDA questionnaire, preferring instead to use their budget to pay for questions to be added to CDC's National Health Interview Survey and Behavioral Risk Factor Surveillance System. Serious collaboration between OAS and CMHS involving the NHSDA did not occur until around 2006.

Officials at ONDCP and SAMHSA felt the NHSDA was not useful to the states and did not provide the data that was needed to manage the block grant program. Some at SAMHSA preferred that the NHSDA be substantially reduced or eliminated, with the funding shifted to CSAT or distributed directly to states to support their own surveys, tailored to their individual needs. Some state officials agreed with this approach. But ONDCP would not allow this to happen. They valued the national data, and favored a further expansion of NHSDA to get state-level data.

Building OAS

With the structure of the new organization in place, HHS next needed to determine staffing. Which staff from the institutes would go to SAMHSA and which would go to NIH? Answers did not come easily.

At issue were two of the branches within NIDA's Division of Epidemiology and Prevention Research (DEPR). I was chief of the Survey and Analysis Branch, which was responsible for NHSDA and parts of

DAWN, TEDS, and NDATUS. There were twelve people in the branch. Seven of the twelve had some NHSDA responsibility.[3] The Systems Analysis Branch, supervised by Betsy Slay, had eleven staff members. They handled data processing and maintained computer systems associated with the surveys.

NIDA leadership was asked how many staff members were needed to manage the data systems. The initial response from NIDA's acting director Richard Millstein was that seven people from DEPR should move to SAMHSA to manage the data systems, including only one assigned to NHSDA: project officer Joe Gustin.[4] NIDA said this would give SAMHSA the resources to undertake project administration, management, and data reporting, while NIDA would retain responsibility for questionnaire design, secondary analysis, and methodological research. ADAMHA acting administrator Elaine Johnson, who would soon become the acting SAMHSA administrator, wisely told NIDA that SAMHSA needed a staffing level sufficient to carry out the full range of survey design and analysis work – all twenty-three of the staff in the two branches. NIDA countered with an offer of eleven staff members from the two branches. SAMHSA rejected this offer as well.

Up to this point, I had no involvement in these negotiations, except to provide the DEPR director with background information on the work that my branch did. I was puzzled by NIDA's initial offer. Did they really think seven people could manage all of these complex projects, including just one person to manage the NHSDA? Were they serious in proposing that NIDA be responsible for questionnaire design on SAMHSA's survey? Perhaps these unrealistic proposals were evidence of the wisdom of moving the surveys out of NIDA. Maybe SAMHSA would create an environment where survey research and participation in the federal statistical system were better understood and valued. Yet, at the same time, NIDA and DEPR leaders were warning staff that SAMHSA would not be a desirable place to work, saying "You don't want to work for those people." Was NIDA looking out for its employees, or were they just trying to maintain staffing and funding levels?

NIDA's early position in these negotiations should have been surprising to HHS officials. Only a few months before, NIDA's memo to Assistant Secretary for Health James Mason asserted that one cause for the NHSDA imputation error was that NIDA did not have enough staff managing the survey and checking the data. At that time, NIDA was "examining whether an increase in staff devoted to NHSDA, and to all the survey and analysis work, is needed." Ironically, this explanation was provided to Mason in a memo, drafted by NIDA and officially from Elaine Johnson.[5] Now during the staffing negotiations, NIDA was telling

Johnson (who was now representing SAMHSA) that SAMSHA would only need one person to manage NHSDA. Johnson knew this was a ridiculous proposal, but was unwilling or unable to reach a satisfactory agreement with NIDA on staffing.

The HHS Office of Management was asked to resolve the issue. They held separate meetings with Betsy Slay and me. I told them both NIDA and SAMHSA would need statistical and epidemiologic staff to carry out their responsibilities, and both agencies should retain and recruit more staff to conduct analysis of the survey data. We discussed the current responsibilities and expertise of each member of my branch.

Based on these discussions, the final recommendations were outlined in a memo to Assistant Secretary for Health Mason. He approved. Seven of the twelve people in my branch, including myself and six others, would move to SAMHSA with the surveys.[6] Three of the seven were from the NHSDA team, and the other four worked on DAWN, TEDS, and NDATUS. All eleven of the people in Betsy's branch would also move to SAMHSA. The memo said SAMHSA should have full responsibility for all aspects of the surveys, and it was crucial to keep the same people doing the survey work, retaining institutional memory and expertise on the surveys. SAMHSA should have adequate staff and resources to effectively manage a strong data collection program, including the analytic capability to identify priorities, redesign the survey as needed, and interpret data. Furthermore, it was recommended that the staff transferred to SAMHSA should continue to have responsibility for secondary data analysis, including publishing research articles, as this is generally regarded as good practice for statistical organizations and would help attract a high caliber of new staff.[7]

With staffing for the data systems set, the final ADAMHA reorganization plan was completed and approved by NIH Director Healy, Acting ADAMHA Administrator Johnson, Assistant Secretary for Health Mason, and HHS Secretary Sullivan. Dr. Daniel Melnick was named acting director of The Office of Applied Studies (OAS). He began to develop an organizational structure for OAS. Also, the loss of staff supporting the NHSDA created an urgent need to recruit. Two new statisticians were on board by January.[8]

New Leaders and Shifting Policies in the Drug War

President Bill Clinton took office on January 20, 1993. His nominee for Director of ONDCP was Lee Brown, former Chief of Police in Houston and police commissioner of New York City. The new HHS Secretary was Dr. Donna Shalala, former Chancellor of the University of

Wisconsin–Madison. At SAMHSA, Elaine Johnson stayed on as acting administrator until the first presidentially appointed administrator, Dr. Nelba Chavez, was approved by the Senate in July 1994.

During the election campaign, Clinton promised to reduce White House staff by 25 percent. He kept this promise, and a big chunk of that reduction came from ONDCP, where half of the staff were political appointees under Bush. The number of positions was cut from 146 to 25. On the other hand, Clinton also elevated the drug czar to cabinet status, recognizing at least symbolically the continuing importance of ONDCP in the drug war. The Clinton Administration shifted the focus of the drug war from overseas programs and domestic law enforcement to treatment.[9] The administration favored a more integrated approach that would deal with the drug problem in conjunction with other social and economic issues, as described in the 1993 Interim Drug Control Strategy:

> We will make drug policy a cornerstone of domestic policy in general and social policy in particular, by acknowledging drug abuse as a public health problem and by linking drug policy to our efforts to grow the economy, to empower communities, to curb youth violence, to preserve families, and to reform health care.[10]

1992 NHSDA: Building a Consistent Timetable and Statistical Integrity

Continuous data collection for NHSDA began in January 1992. Each quarter, interviewers were given a new sample of addresses to contact, keeping them busy for three months.[11] Updates to the questionnaire and data collection procedures were implemented each January, and all interviewers attended training sessions during the first week of January.

The sample design for 1992 oversampled the same six metropolitan areas, as recommended by Jim Burke, Chairman of the Partnership for a Drug Free America. The final sample size was 28,832 and the response rate was 83 percent. Questionnaire changes between 1991 and 1992 were minor. The script for the interviewer's introductory comments to the respondent was expanded to more clearly explain the reasons for the study and the importance of confidentiality. Crack questions were removed from the "Cocaine"section and placed in a new "Crack" module. Questions on needle use were shifted from the "Drugs" module to a new "Needle Use" module. Questions about how the drug was obtained were added to the Marijuana, Cocaine, and Crack modules. A question about how much money was spent on marijuana was added to the Marijuana module. Five questions about the last treatment received (when, where, reasons, outcome, and how paid for) were added to the Treatment module. The survey was now housed in a new agency, with new priorities. Change was inevitable. But, of course, that was nothing new. Change had become the norm for the drug survey. On the other hand, the establishment of annual surveys with continuous data collection created a new opportunity for stability. Now we could develop a timetable for processing the data and reporting results, and it would be the same every year. Data collection would be completed every December, and results would be released about seven months later. Just as in prior years, there were many processes that would occur during that seven-month period. Some had to be done sequentially, for example, editing before imputation, and imputation before final table production. Other tasks could be done simultaneously, such as computing weights while imputation was being done. Table shells (the detailed outline of what each table would show) had to be designed early in the timeline, so the tables were programmed and ready for the estimates to be inserted as soon as the weights and final imputed data were ready. There was a lot going on, but at least now the schedule was somewhat predictable, which was helpful in planning. But a new complication was monitoring the simultaneous work on three different surveys at different stages. A complete survey cycle, from design through data file delivery, was more than three years long. For example, in January 1993 the initial data checking and editing of the 1992 data began, as well as the development of table shells. At the same time, several hundred interviewers were trained and began knocking on doors across the country for the 1993 survey. Thirdly, sample construction and questionnaire development for the 1994 survey were well underway. The contractor had a large staff handling all these facets of the surveys. SAMHSA had a team of five statisticians managing the project, and NHSDA was not their only responsibility. This was far too few to adequately monitor all of the contractor's work. It was also risky, because with the fixed, continuous data collection schedule, late completion of tasks and delays in clearances and approvals would be disastrous and costly.

The period between table production and a late summer release of a report was not long enough to review the results, check for errors, and compile the data into a comprehensive summary report of the findings that included a description of the survey methodology. This type of report, called Main Findings, had been compiled for each previous National Survey. But the report was never finished in time for the press conferences announcing the results. Instead, the survey team prepared data tables, briefings, and a few pages of highlights prior to the data release. These materials were used by the press officers, analysts, agency leaders, and speechwriters to develop their own interpretations and statements about the results. That was all that was available to reporters and the public during and immediately after the press conferences.

We knew that it was important to have a comprehensive, printed report available to the public the same day that the results were released. We felt that this report should be produced by our office, providing background information on the survey design, caveats, sampling and non-sampling error, and appropriate interpretation of survey results. It should provide an objective assessment of the survey estimates and interpretation of the results, independent of political influence or policy preferences.[12] This approach would have multiple benefits. The results would be more likely to be perceived as credible if the data producers fully disclose how the estimates were made, including explanations of the accuracy and caveats associated with the data. This kind of openness and full disclosure contributes to "statistical integrity," which refers to the scientific rigor that gives credibility to the data and its producers. In addition, the report would give the media and others interested in the findings a document to read to better understand the data source, and to see more of the results than what was selectively highlighted in the press release and speeches. Finally, prior to the release, the report could be used by agency leaders, speechwriters, and press officers as the basis of their remarks and handouts at the press conference. The hope was that this would result in a more accurate presentation of the results to the public – thus protecting the government officials from making unsupported claims that could be criticized by the media or by political opponents. Statistical integrity protects the drug czar just as it does the survey managers.

Before the report on the 1992 survey could be written, an unusual result had to be investigated. An initial check of the first six months of data revealed questionable declines in substance use among blacks between 1991 and 1992. Once a person is a lifetime user, they remain so the rest of their life, so a large decline in this rate over a short period of time is unrealistic. We established a Peer Review Committee consisting of substance abuse researchers, survey design experts, and health statisticians to review the data and advise us on these results.[13] After several meetings, the committee concluded "the observed differences between 1991 and 1992 cannot be explained by a single factor," and the unusual decline among blacks "is an example of what can occasionally occur in survey estimates, particularly when a large number of estimates are generated and comparisons are made." After a careful review of the survey methodology, they stated "the design and procedures for sampling, weighting, editing, and imputing the survey results are statistically sound." They recommended the estimates for 1992 be released by SAMHSA, along with footnotes or caveats indicating caution in comparing prior survey estimates for blacks.

The report on the findings of the 1992 survey was called "Preliminary Estimates from the 1992 National Household Survey on Drug Abuse." It was report number three in a new OAS publication series, called "Advance Reports." The brainchild of OAS acting director Dan Melnick, the Advance Reports were produced entirely in house, by OAS staff. The NHSDA report was sixty-three pages long. It included sixteen pages of brief bulleted findings, ten pages of appendices explaining the survey methodology, and thirty pages of tables. "Preliminary" in the title of the report reflected a concern that errors might be found after the release, as had occurred with the 1991 estimates. The appendix stated that "Further analyses of the 1992 NHSDA data and evaluation of the estimation procedures is ongoing, and may result in revisions in later data releases." Consistent with the policy of full disclosure of all known caveats, the appendix included a discussion of the Peer Review Committee. Another new feature of the report was a discussion of other sources of data on substance use. This helped to put the NHSDA results in the context of findings from other studies, comparing results and explaining how different methods employed in other studies might cause conflicting findings. It gave the report a more comprehensive approach, rather than a narrow focus on one survey. The format and content of the report, although developed quickly, turned out to be well received and long-lasting. Twenty years later, National Findings reports looked very much the same, except they were longer, more colorful, and available on the web.

A key characteristic of the early Advance Reports was that they were not treated as regular agency publications, which had to go through elaborate departmental clearances (and potential delays) and printing by the Government Printing Office. Bypassing those bureaucratic uncertainties helped make it possible to have the report ready for distribution on the day of the press conference.

Results were released at a press conference at HHS headquarters, June 23, 1993, less than six months after the end of data collection. Overall, the survey showed continuing declines in overall illicit drug use. Frequent or weekly cocaine use, however, remained unchanged between 1991 and 1992, at about 640,000 users. "In fact, no significant change has occurred in this number since it was first estimated in 1985," HHS Secretary Donna Shalala stated. Drug Czar Lee Brown did not participate, but issued a press release the same day that contradicted Shalala. Brown said the NHSDA was imperfect, and "tells only part of the story. From other data – reflected in the Household Survey's reported rise in frequent cocaine use – we know that hard core drug use continues unabated."

Notes

1 Institute of Medicine, *Research and Service Programs*, 70–88.

2 John Walters, Chief of Staff, ONDCP memorandum to Mark Barnes, Counsel to the Secretary for Drug Abuse Policy, HHS. July 16, 1991.

3 Marc Brodsky, Joe Gfroerer, Janet Greenblatt, Joe Gustin, Lana Harrison, Art Hughes, and Andrea Kopstein.

4 Mr. Gustin had been alternate project officer since 1989, managing the day to day operations and closely monitoring budget and costs. Anticipating the reorganization, DEPR elevated him to project officer, expecting that if the survey was transferred to SAMHSA, he would move to SAMHSA with the survey to maintain oversight continuity while the branch chief (myself) would remain at DEPR.

5 Elaine Johnson, memorandum to James Mason, June 12, 1992.

6 My SAMHSA appointment was a detail, with an option to return to the NIDA branch chief position after one year. I later opted to stay at SAMHSA. Other NSDUH staff that were transferred to SAMHSA were Joe Gustin and Janet Greenblatt.

7 Anthony Itteilag, Deputy Assistant Secretary for Health Management Operations, memorandum to James Mason, August 17, 1992.

8 Joan Epstein, formerly of Census Bureau, and Doug Wright, formerly of National Center for Education Statistics.

9 Whitford and Yates, *Presidential Rhetoric,* 66.

10 ONDCP, *National Drug Control Strategy, 1993*, 5.

11 Although each quarterly sample is representative of the US population, reports and analyses are usually based on the combined four quarters of data for the year.

12 National Research Council, *Principles and Practices*, 6–8.

13 Members of the Peer Review Committee were James Massey (NCHS; Chair), Marc Brodsky (NIDA), Joe Gfroerer (OAS), Tom Harford (NIAAA), Lana Harrison (NIDA), Marilyn Henderson (CMHS), Dale Hitchcock (HHS), Ron Manderscheid (CMHS), Nancy Pearce (HHS), Beatrice Rouse (OAS), and Ron Wilson (NCHS).

7 Rising Youth Drug Use in the 1990s

After a period of decline in youth illicit drug use during the 1980s, evidence of a reversal in the trend began to appear in the early 1990s. The larger NHSDA sample size in 1991 and continuous data collection in 1992 occurred just as the increase in drug use was starting. The benefits of the survey changes were immediately evident, and the NHSDA was established as a critical tool, along with the Monitoring the Future study (MTF), for tracking youth trends. Under HHS Secretary Donna Shalala, there was a strong reliance on survey data to define the problem and develop solutions.

Shalala was adept in working with data, and interested in research. Each year, before the release of the data, she asked SAMHSA for a full briefing on the results, presented by myself and the OAS director. She usually invited her senior staff as well as representatives of other HHS agencies and sometimes ONDCP. There was always a lively and informative discussion at these briefings, led by Shalala. She convened separate meetings to discuss other data issues, such as why the increase in youth drug use was occurring and our proposed methods for getting state estimates from NHSDA. She was creative in scheduling her press conferences, holding them at local youth facilities such as a high school library and a Boys & Girls Club to convey directly to teens the message that drug use was dangerous.

1993 NHSDA

The sample design for 1993 was similar to the 1992 design, including oversamples in the same six metropolitan areas. The sample size was 26,489, and the response rate was 79 percent. A sampling refinement was introduced, in response to ONDCP's request to oversample drug users. After studying the possibilities, we decided to try oversampling cigarette smokers, based on the fact that a high proportion of illicit drug users also smoke. To do this, an item was added to the household screening questionnaire, asking whether or not each household member between the ages of 18 and 34 was a current smoker. Smokers in that age range in the household were given a greater probability of selection for the survey interview. The 1993 questionnaire was nearly identical to the 1992 questionnaire. The changes included the addition of new questions on heroin snorting, sniffing, and smoking, and reformatting of the questions on the use of drugs in combination with alcohol, to reduce reporting errors caused by the poor question construction discussed in Chapter 5.[1]

Results were released in Advance Report Number 7, at a press conference at HHS headquarters on July 20, 1994. HHS Secretary Donna Shalala led the event. She summarized the findings as showing no changes in youth drug use between 1992 and 1993. This was not good news, since the rates had been declining between 1979 and 1992. Reflecting the administration policy of greater emphasis on treatment, she said "We must enhance and expand treatment options for so-called chronic hard-core drug abusers. The President has made three significant proposals to expand drug treatment: one is his 1995 budget, a second is his health care reform bill, and a third for treatment in the criminal justice system as part of the crime bill. We need to enact all three." The emphasis on expanding treatment was supported by a recently released study that concluded treatment was seven times more effective than domestic law enforcement at reducing cocaine consumption.[2]

The speakers at the press conference did not mention any resurgence in drug use among youths. But evidence of an increase began emerging more than a year earlier, in the 1992 MTF data. Released in April 1993, that survey reported increases in use among eighth graders between 1991 and 1992. Increases among eighth, tenth and twelfth graders were seen in the 1993 MTF results released in January 1994. There was also evidence in the 1993 NHSDA data, but it was reported cautiously in the Advance Report due to the lack of statistical significance in past month use, a key tracking measure: "The rate of marijuana use for 12–17 year olds was 4.0 percent in 1992 and 4.9 percent in 1993. While this is not a statistically significant change, it is still noteworthy given the recently reported increase among high school students in the Monitoring the Future Study. Furthermore, the NHSDA did find a significant increase in *past year* marijuana use among youths (8.1 percent in 1992 and 10.1 percent in 1993)."

1994 NHSDA: Tracking Trends during a Survey Redesign

Due to a reduced budget, cost reductions in the NSDUH were needed. Cutting back on the oversampling of metropolitan areas was a prime candidate for cutting costs. The extra sample in the six cities did little to improve the precision of national estimates. We convened an expert panel that included representatives from state and local agencies.[3] Although consultants from New York and Miami said the estimates for their cities have been useful, the panel generally agreed the most logical way to reduce the survey cost was to drop the metropolitan area oversampling. It would reduce the estimated cost for the 1994 survey from $12 million to $8 million, with only a small impact on the precision of national estimates. We followed this recommendation, and in early 1993, began a new contract with RTI to conduct the 1994–96 surveys. Each survey would employ a national sample of about 18,000 respondents. In addition, the Department of Agriculture provided funding to increase the rural area sample by 1,000 interviews in 1994.

The NHSDA team made changes to the questionnaire and data processing procedures in 1994. These changes were based on the results of the methodological studies started in 1989. Originally we had planned to introduce the new questionnaire in 1992, but we needed more time to develop and test it. The major features of the new questionnaire were:

- Questions about tobacco use were asked using a self-administered answer sheet. In prior survey years, these questions were interviewer-administered.
- A priority set of demographic and drug use items was designated as "core" and these items were placed at the beginning of the interview. The idea was that these core items would remain unchanged in future years to maintain comparability. The remaining sections were designated as "non-core" and, in theory at least, could be revised, with questions being dropped and added as the data needs changed.
- Questions were reworded to create consistent wording across sections and to improve clarity and ease of understanding.
- The order of the self-administered answer sheets was changed.
- A new, clearer definition of nonmedical use of prescription drugs was introduced: use of a prescription drug "when it was not prescribed for you, or only for the feeling or experience it caused."
- A calendar was given to respondents as an aid to help them answer questions on past thirty days and past twelve months.

Funding for the survey was also received from the Department of Labor, for the inclusion of a "Workplace Experiences" module. This module, placed near the end of the NSDUH questionnaire, asked employed respondents about the substance use policies and practices of their employer, including drug testing and employee assistance programs. Although the funding for the module was just for 1994, SAMHSA retained most of the workplace questions on the survey in subsequent years, primarily due to the urging of SAMHSA's Division of Workplace Programs.

The revised editing procedures were intended to prevent overestimation that may have occurred with the previous editing method. Under the old editing rules, *any* mention of use of a drug during a specific time period *anywhere* in the questionnaire was enough to code a respondent as a user, regardless of how many denials of use or unanswered questions there were. The revised rule was more conservative in coding a respondent to be a drug user when his/her responses were inconsistent (i.e., they report use on one or more questions, but nonuse on other questions asking about the same time period). The new method was referred to as an "adjusted single-mention core" procedure. A respondent could be coded as a user based on a single mention of use, but only if that mention was in the core section. And in cases where the respondent indicated "never used" on the recency question, and all but one of the other questions about drug use were denials, the respondent would *not* be coded as a lifetime user.

We anticipated the questionnaire and editing changes would result in estimates that were not comparable to estimates from prior years. So we increased the total sample size and administered the 1993 questionnaire to a portion of the 1994 sample, referring to it as the 1994-A questionnaire. The new questionnaire was 1994-B. One-fifth of the sample was randomly assigned to sample A.[4] The final sample sizes were 4,372 for sample A and 17,809 for sample B. Response rates were 77 and 78 percent, respectively. Old editing rules were applied to the A sample. Advance Report 10 included two sets of estimates – one from each sample. Sample A estimates were used in describing trends from 1979 to 1994, while sample B estimates were used to describe the patterns of use in 1994.

The first sentence of the HHS press release announcing the results of the 1994 survey came directly from the OAS report: "Marijuana use among 12–17 year olds nearly doubled from 1992 to 1994, though it remains far below the peak reached in 1979, according to the 1994 National Household Survey on Drug Abuse." Both HHS Secretary Shalala and drug czar Brown highlighted this trend in their remarks at the press conference. The focus on youth was accentuated by the setting: the library of Woodrow Wilson High School in Washington, DC. It was

September 12, 1995, the start of a new school year. Students, teachers, and reporters huddled around Brown and Shalala as they spoke about the data and displayed new public education materials designed to help young people resist drugs. Dr. Nelba Chavez, the recently appointed administrator of SAMHSA, also spoke. Her closing statement nicely summed up the role for the survey within her services-oriented agency: "The Household Survey is one important part of SAMHSA's leadership responsibility. It ensures that we remember the past, it tells us where we are and helps us target our resources in order to more successfully chart the future."

Attacks on Clinton's Drug Policy

The NHSDA and MTF data, showing increasing drug use among youths, became fodder for the Republican attacks on President Clinton, in the months leading up to the 1996 presidential election. As was the case four years earlier, it was the Chairman of the Senate Committee on the Judiciary leading the critics. But this time, the chairman was Republican Orrin Hatch. In December 1995, he issued "Losing Ground Against Drugs, a Report on Increasing Illicit Drug Use and National Drug Policy." The report stated:

> Federal drug policy is at a crossroads. Ineffectual leadership and failed federal policies have combined with ambiguous cultural messages to generate changing attitudes among our young people and sharp increases in youthful drug use.

The Hatch report criticized the Clinton administration's policy that de-emphasized law enforcement, shifting resources away from interdiction and towards treatment of hard-core drug users. The report said there was no evidence of greater availability of treatment, and noted continuing increases in hard core drug use.[5] Because of Clinton's cuts in interdiction funding, Hatch claimed, "Illicit drugs are now available in greater quantities, at higher purity, and at lower prices than ever before." In reality, with Republicans in the majority in both houses, Congress rejected most of the Clinton proposals to shift funding from interdiction to treatment. As a result, there was a decrease between 1993 and 1997 in the proportion of the federal drug control budget that was allocated to demand reduction.[6]

A few months after the Hatch report came out, a Task Force on National Drug Policy, led by Senate Majority Leader Robert Dole (the eventual Republican Presidential nominee) and House Speaker

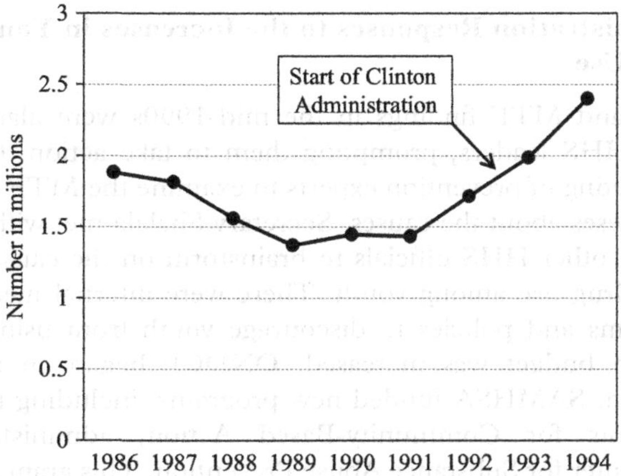

Figure 7.1 Number of marijuana initiates, by year.

Newt Gingrich, released a report called "Setting the Course: A National Drug Strategy." The report began "The facts are simple. After more than a decade of decline, teenage drug use is on the rise. Dramatically." The report claimed the Clinton administration downplayed the drug issue, de-emphasizing the message that drug use was wrong. "The result was to replace 'Just Say No' with 'Just Say Nothing.'" The report even cited Representative Charles Rangel, Democratic Chairman of the House Select Committee on Narcotics, saying "I have been in Congress for over two decades, and I have never, never, never found any administration that has been so silent on this great challenge to the American people." But the report also blamed parents, rock stars, television, and movies. Parents talked less to their kids about the dangers of drug use. Drug use was presented in a favorable light on MTV. Musicians and TV shows like *Saturday Night Live* depicted it as harmless fun. These societal changes probably had more impact on youth drug use trends than any shift in government policies. Survey data did not support the claim that the upward trend in teen drug use was caused by Clinton administration policies. Increases were initially found in the 1992 MTF data and were even more pronounced in the 1993 MTF data. NHSDA estimates of marijuana initiation in Figure 7.1 also show increases beginning in 1992.[7] Clinton did not take office until January 1993. Clinton officials would make this point when the 1995 NHSDA results were released, about ten weeks before election day.

Administration Responses to the Increases in Youth Drug Use

The NHSDA and MTF findings in the mid-1990s were alarming to ONDCP and HHS leaders, prompting them to take action. ONDCP convened a meeting of prevention experts to examine the MTF data and develop hypotheses about the causes. Secretary Shalala met with NIDA, SAMHSA, and other HHS officials to brainstorm on the causes of the resurgence of drug use among youth. There were internal meetings to identify programs and policies to discourage youth from using drugs. The prevention budget was increased. ONDCP began an intensive media campaign. SAMHSA funded new programs, including the State Incentive Grants for Community-Based Action, administered by SAMHSA's Center for Substance Abuse Prevention. This grant program helped identify successful prevention approaches and facilitated their implementation in many communities. New prevention-related pamphlets and other resources were developed. The increase in youth drug use was also a key factor behind the Secretary's initiative to expand the sample size of the NHSDA to make state estimates possible. Eventually implemented in 1999, this sample redesign is discussed in Chapter 9.

1995 NHSDA: Creating a Bridge to the Past

With the new questionnaire in place starting in 1994, there was not much need or desire to make changes in 1995. The core sections were not altered. In the non-core part of the interview, new questions on the perceived need for substance use treatment in the past twelve months were added. Also, there was a new series of questions on arrests by specific type of crime and types of crime committed in the past twelve months, to obtain data needed to study the connection between substance use and crime. The sample size was 17,747. The response rate was 81 percent.

The results were published in Advance Report Number 18, "Preliminary Estimates from the 1995 National Household Survey on Drug Abuse." As is usually the case, comparisons between this year and last year were of great interest. The survey was well positioned to identify changes in drug use between 1994 and 1995. The 1995 and the 1994-B samples each had nearly 18,000 interviews, with the same questionnaire and the same sample design. But what about long-term trends? We needed to be able to compare the 1995 estimates with 1979 (the peak year for illicit drug use rates among youths) and with 1992 (the low point for youth drug use). But the questionnaire and editing procedures were different across the different years. And for cigarette use, the mode of data collection changed from interviewer-administered to self-administered in 1994.

The split sample design in 1994 provided a way to "bridge" the old and new data. Of course, this was no accident. It was the purpose of the random assignment of respondents to the A and B samples. Our hope was that we could construct adjustment factors that would be applied to the 1993 and earlier estimates, to create a consistent long-term trend. We were not sure it would work well, and we accepted that it would be crude. But we had to try. We asked RTI to explore the possibilities. Dr. Ralph Folsom led this effort. The key factor underlying the study was that the 1994-A and B samples were simultaneous representative samples of the same population. It was essentially a controlled experiment. The experimental condition (the factor that was altered) was the questionnaire and its associated editing rules. But our goal was not just to test whether the new questionnaire had a significant effect on drug use estimates. We also wanted to quantify this effect. Did the new method result in a 20 percent higher rate of use? Or a 50 percent lower rate? And we wanted to measure this separately for each drug, for each time period (lifetime, past year, past month) and for different demographic groups. We discovered that the 1994-A sample of about 4,400 cases, was too small to be able to determine differences at that level of detail. Our adjustments would have to be very crude. We did not have a large enough sample to accurately estimate the effects of the new questionnaire for low-prevalence behaviors such as past month cocaine use, and we could not break out differences by many demographic groups. But it was also clear that the new questionnaire and editing did have an impact on estimates. The shift from open interview to self-administration for the tobacco questions resulted in higher rates of cigarette use, especially for youths. The 1994-A sample estimated that 9.8 percent of youths smoked within the past month, about the same as in 1993 (9.6 percent). Using the private answer sheets, the 1994-B sample produced an estimate of 18.9 percent. The answer sheets also produced higher smoking rates for adults. The higher rates were partly due to the increased privacy, which made respondents more willing to disclose their smoking behavior. It was also due to the way respondents were routed through the questionnaire (the skip patterns). In the old interviewer-administered questionnaire, if a respondent initially said they had never used cigarettes, they were routed to the next module, on alcohol use. There were no more opportunities to report their smoking. But the self-administered answer sheets require everyone, regardless of whether or not they ever used the substance, to answer all questions. An initial "never used" response could be edited to "used" if the answer to a later question indicated some use. The effects of the new questionnaire were not nearly as large for other substances, which had all been administered via the answer sheets in the previous surveys. In general, differences between 1994-A and 1994-B estimates were small. Most estimates were slightly lower with the new questionnaire. This was not surprising, since the new editing rule was more conservative. It was less likely to "create" a drug user based on sparse evidence.[8]

RTI was successful in developing some statistical adjustments that appeared to be reasonably accurate. We decided to limit the adjustment to a specified subset of measures and populations. We chose measures that were of most interest to data users. Adjusted estimates were created for the years 1979 to 1993, and entered into tables examining trends across time along with the 1994-B and 1995 estimates. Most of the original 1979–93 substance use prevalence estimates were made smaller by the adjustment, with the exceptions of cigarette and smokeless tobacco use. Sampling errors were also recalculated for all of the adjusted estimates, accounting for not only the original sampling in the 1979–93 surveys, but also the sampling error due to the 1994 sample from which the adjustment factors were estimated.[9]

Pre-Release Controls and a Leak of the 1995 Survey Findings

New leaders were in place for the 1995 data release. In the summer of 1996, Dr. Donald Goldstone was appointed as the first permanent director of SAMHSA's Office of Applied Studies. Don had many years of experience in managing health survey research programs in HHS, and had a good understanding of how large government surveys operate. He eventually became quite knowledgeable of the NHSDA data collection process and familiar with the data. Also that summer, General Barry McCaffrey took over as Director of ONDCP.

By 1996, the annual NHSDA data release procedure had become more predictable. With a consistent data collection and processing schedule, we could anticipate a late summer release each year. An OAS publication at the time of the release was expected. This consistency made it easier to efficiently manage the production of the estimates, and prepare a high quality report. OAS survey staff and RTI could plan ahead, organize resources, set timetables, and track progress towards an ultimate goal of having the report ready by July. But there was still uncertainty about the actual release date. A regularly scheduled data release would have been beneficial to SAMHSA, HHS, and ONDCP.

When agencies set a release schedule and stick to it, the media and the public have more trust in the results. There is less of an opportunity for the manipulation of the timing of the release of results for political purposes.

But for NHSDA in the 1990s, a regular release date was not established. Here is how the release occurred in a typical year. Once RTI finished editing, imputation, and weighting of the data, they would begin sending tables to me. This would occur around April. The NHSDA team would begin checking the tables, looking for unusual patterns that could signal an error or an important new finding. Once data issues were investigated and resolved, we wrote the report. We submitted the report to the OAS director, who then reviewed it and gave us his changes. Revisions were made and the report was given to the SAMHSA communications director. Mark Weber took on this role just before the release of the 1995 data.[10] He rarely suggested any changes to the report, and he set up briefings for the SAMHSA administrator and senior staff, the HHS secretary, the drug czar, the NIDA director, and sometimes outside constituent groups or Congressional staff and members. These briefings typically consisted of a thirty-minute presentation (by me or the OAS director) of about forty slides showing line graphs and bar charts summarizing the key findings of the survey, followed by a discussion of the results and plans for release. At some point, HHS and the White House would agree on a release date, venue, and participants. The availability of the key participants (drug czar, HHS secretary, and SAMHSA administrator) was a key factor in determining the date.

During the several weeks between the time the report was completed and the press conference, "close hold" was the rule. The data were treated as if they were top secret. They were shared only with a select few who absolutely needed to know. Those who were invited to briefings were warned not to share what they heard with anyone outside the room, and usually there were no paper copies with data handed out. Sometimes when printed copies of the report were shared with selected officials prior to the release, each copy was numbered and tracked. One time a stray copy was found in the ladies room at HHS headquarters. The number identified the perpetrator, and she was promptly notified and scolded. Why such secrecy? I suspect it was mainly because it was important to maximize the impact of the press conference. When the Secretary of HHS announced the results, it had to be news. Not old news. Tight controls on sharing of the results were also due to the chain of command and bureaucratic protocols. The report would not be shared with the Secretary's office until it was first shared with the SAMHSA administrator. And it would not go to ONDCP until the HHS Secretary had seen it. One year, weeks before the release, SAMHSA Administrator Nelba Chavez received a call from Secretary Shalala. Shalala was upset because President Clinton had just told her about the good news he had heard from Barry McCaffrey. The new survey data showed a decline in youth drug use. This was news to Shalala. Had SAMHSA shared the survey results with ONDCP before she had seen them? She called Chavez to get to the bottom of the incident. At first, Nelba thought it meant trouble for her and her staff, but she checked with her communications director Mark Weber who told her that SAMHSA did not break any protocols. The report had been sent to Shalala's staff before it went to ONDCP, but staff had not yet had a chance to tell her about it.

The handling of the drug war by the federal government was often an issue in national politics, especially in the 1980s and 1990s. Leaks of the survey results to the media or Congress were a concern. It happened in 1996 with the 1995 survey results. Three months before President Clinton would face Senator Robert Dole in the election, and two weeks before the HHS press conference to release the survey results, *US News and World Report* published some of the 1995 estimates from "an advance copy" they obtained.[11] At the same time, the *Washington Times* published an editorial, highly critical of the Clinton administration's drug abuse strategy:

> Rarely in the public policy game has one administration bequeathed to another such a win-win situation as the national drug-control policy that the Bush administration handed off to President Clinton. Rarer still are those instances when the incoming administration has so utterly mangled and mismanaged such promising opportunities–as the Clinton administration has done in its failure-plagued drug abuse strategy.
>
> Even the administration's new drug czar, Barry McCaffrey, minces no words in his acknowledgement of the disaster he has inherited. "The bad news is kids are using drugs again," he admitted in late June. "Drug use among children–adolescents–is now within emergency levels."
>
> That acknowledgement preceded the latest findings of the National Household Survey on Drug Abuse conducted under the auspices of the US Substance Abuse and Mental Health Services Administration. The administration has not yet made public the explosive data from 1995, but the *Washington Times'* editorial page has obtained a copy. And the "bad news" about the "emergency levels" of adolescent drug use decried by Mr. McCaffrey has just gotten significantly worse. So depressing are the numbers from the 1995 drug survey, it's no wonder the Clinton administration didn't want it to become public on the eve of the Republican Convention.[12]

The Republican party nominated Robert Dole to run for president at their national convention in San Diego, August 12–15, 1996. Drug abuse was not a major focus at the convention. But after the convention, a Dole

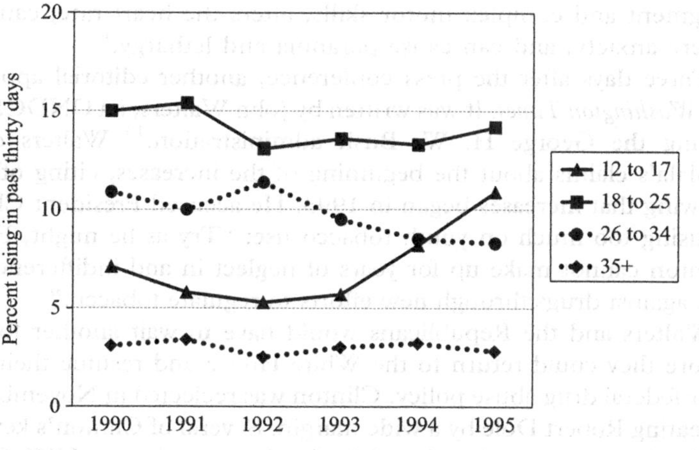

Figure 7.2 Any illicit drug use in the past thirty days, by age group and year.

campaign ad highlighted the doubling of the rate of youth drug use during the Clinton administration, saying that "Clinton never took the drug crisis seriously" and that "America deserves better." The thirty-second television spot showed cleverly edited clips of Clinton, appearing to show him saying that he wished he had inhaled marijuana, and then saying he "wished he had never done any of that . . . it was wrong."

When HHS officially released the NSDUH results on August 20, the numbers were the same as the *Washington Times* had reported, but the description was less dire. There was even some good news to report. Overall drug use remained flat and far below the peak levels of the late 1970s; there was a 35 percent decline in the number of occasional cocaine users since 1991. But Figure 7.2 shows that past month illicit drug use among youth did go up, from 8.2 percent in 1994 to 10.9 percent in 1995. The rate was more than twice what it had been in 1992.

In discussing this continuing increase, Secretary Shalala said she was often asked when these increases began. "The answer, according to the science, is 1991 and 1992," she said. Shalala described it as "a multi-year trend that began before this Administration came to Washington," and went on to say "This President has displayed real leadership and put in place the most comprehensive drug control strategy our nation has ever seen." SAMHSA Administrator Chavez focused on the health effects of marijuana, warning that "Marijuana use damages short-term memory, distorts perception, impairs judgment and complex motor skills, alters the heart rate, can lead to severe anxiety, and can cause paranoia and lethargy."

Three days after the press conference, another editorial appeared in the *Washington Times*. It was written by John Walters, an ONDCP official during the George H. W. Bush administration.[13] Walters disputed Shalala's claims about the beginning of the increases, citing other data showing that increases began in 1993. He accused President Clinton of focusing too much on youth tobacco use: "Try as he might, President Clinton cannot make up for years of neglect in and indifference to the war against drugs through new efforts to regulate tobacco."

Walters and the Republicans would have to wait another four years before they could return to the White House and resume their control over federal drug abuse policy. Clinton was reelected in November 1996, defeating Robert Dole by a wide margin. Several of Clinton's key leaders in the drug war stayed in their jobs for the second term: HHS Secretary Donna Shalala, SAMHSA Administrator Nelba Chavez, and ONDCP Director Barry McCaffrey.

States Legalize Medical Use of Marijuana

Although neither Clinton nor Dole advocated any relaxation of drug laws during the campaign, marijuana and other drug legalization was an issue in the 1996 election. On the ballots in California and Arizona were new laws that legalized medical use of marijuana.[14] California's Proposition 215 was the most concerning to the Clinton administration. It would make it legal to possess and cultivate marijuana for personal use if a doctor recommended it as a course of treatment. Before the election, McCaffrey actively campaigned against Proposition 215. He made several trips to California, giving speeches warning California voters that passage of the law would lead to increasing drug abuse, and that marijuana had no medical value. He threatened that if Proposition 215 passed, he would have federal agents arrest physicians that recommended cannabis to patients. Administration officials continued to speak out against the law after its approval by California voters. "Federal law-enforcement provisions remain in effect. Nothing has changed," McCaffrey stated.[15]

The Arizona law (Proposition 200) approved marijuana and other drugs as safe and effective medicines. But flaws in the way it was written and passed led to its subsequent dismantling by the Arizona legislature. Nevertheless, passage of the two propositions in November 1996 and the expectation that more states would follow added a new urgency to the idea of expanding the NHSDA to get state-level data.

1996 NHSDA

Core sections of the 1996 questionnaire remained unchanged from 1995. In the non-core part of the interview, new modules were added with funding support from outside agencies. The Centers for Disease Control (CDC) wanted data on sexual behavior (see discussion in Chapter 8), and the National Highway Traffic Safety Administration (NHTSA) was interested in driving under the influence of drugs and alcohol. Interagency agreements had been signed prior to data collection, to transfer funding from CDC and NHTSA to SAMHSA. SAMHSA then allocated these funds to the NHSDA contract.

The within-household sampling procedure was simplified in 1996, reverting back to the screening and sampling process that had been used in 1992. A detailed assessment of the impact of the oversampling of current smokers since 1993 found that the gains in precision were small, and only for certain drug variables. For other drug variables, the impact on precision was detrimental. Household screeners would no longer ask about smoking status of household members. The sample size for the 1996 survey was 18,269. The response rate was 79 percent.

The 1996 results were released without a press conference. On August 6, 1997, HHS simply issued a press notice containing a summary of key findings, along with a "Fact Sheet" describing the president's strategy for addressing drug abuse and HHS' role and activities.

The survey estimated that in 1996 there were about 13 million current illicit drug users aged 12 and older. The number and rate were not significantly different than in 1995. The rate of use among youths declined between 1995 and 1996, but increased among young adults age 18–25. There were increases in heroin initiation and past month use since 1992. Most of this increase was attributed to young people, under age 26, smoking, snorting, and sniffing heroin.

The SAMHSA report of the results was formatted exactly as the reports for previous surveys (1992–1995) had been. However, it was published in the new OAS report series, started in 1997 by OAS Director Donald Goldstone. Reports in this series had colorful covers that distinguished the different data sources of each report: Red for NHSDA (H-Series), blue for DAWN (D-Series), green for services data (S-Series), purple for special analytic studies (A-Series), and yellow for methodological reports (M-Series). There were no more "Advance Reports." The NHSDA reports would now go through regular HHS clearance procedures.[16]

1997 NHSDA: Expanded Samples in California and Arizona

The 1997 questionnaire included new items on the use of cigars, and on reasons for using, not using, or quitting use of marijuana and cocaine. A new module called "Youth Experiences" covered a variety of issues that research had shown were associated with substance abuse. These "risk and protective factors" included school grades, participation in activities, relationship with parents, fighting, and stealing. The modules on sexual behaviors and driving behaviors were dropped.

Just before the 1997 data collection was about to begin, SAMHSA was asked by HHS if the NHSDA samples in California and Arizona could be expanded so the impact of the two newly passed laws legalizing medical use of marijuana could be assessed. There was not enough time to accomplish this before the start of data collection in January, but with NHSDA's quarterly sample design, we were able to oversample in the two states starting in the second quarter. The biggest challenge would be to recruit and train more interviewers in California and Arizona and have them ready to knock on doors by April 1. That challenge was successfully met. The extra sampling resulted in sample sizes of 4,360 in California and 4,415 in Arizona, for a total national sample of 24,505. The response rate was 78 percent.

Analysis of the data from the two oversampled states required some special procedures, because the oversample was only for April through December. Separate three-quarter weights were computed for the entire sample, so that estimates representative of California, Arizona, and the United States could be generated from the quarter's 2–4 sample. For comparisons of drug use prevalence between the two states and the remainder of the country, special tables were developed. For most national tables that did not break out data for the two states, the regular twelve-month weights were used. Thus, there were two different sets of national estimates in the reports. It turned out there were only very small differences between the nine-month and the twelve-month national estimates. Nevertheless, we decided it was appropriate to only use the nine-month national estimates in any state-level analysis.

Results were released during a difficult period in the Clinton administration. Unrelated to drug abuse policy or statistics, the news in August 1998 was dominated by the investigation of the president's alleged affair with White House intern Monica Lewinsky. Clinton had consistently denied the allegations, and had been strongly defended by several of his senior officials, including HHS Secretary Shalala. There was anger and disappointment among administration officials when President Clinton announced on August 16 that the allegations were true. The next day, it seemed like some of that anger was unleashed on Don Goldstone and me. Our briefing for Shalala on the 1997 NHSDA results was on the morning of August 17. Every meeting and briefing we had had with her about the survey until then had been cordial, with lively, insightful discussion of the findings. But this one was different. She questioned the accuracy of some of the estimates and my explanation of the methods we used. She was obviously less engaged in the presentation and less interested in the findings than in prior briefings. Simply put, she was in a bad mood. We quickly recognized her demeanor and tiptoed through the session, carefully avoiding saying anything that might trigger more criticisms or skepticism about the survey. Of course we did not take it personally.

Four days later at the press conference, there was no sign of ire and no mention of the president's troubles. The focus was on drug use. The administration may have welcomed the release of the findings more as a distraction from the Lewinsky affair than as an important milestone in the war on drugs. The survey showed that increases in youth marijuana use continued, with the rate of past month use rising from 7.1 to 9.4 percent between 1996 and 1997. Announcing the results, Secretary Shalala highlighted parents' role in keeping kids off drugs. She mentioned the president's National Drug Control Strategy and the administration's plans to form stronger partnerships with states and communities, "and if Congress provides the necessary funding, the National Household Survey will be expanded in 1999 to provide information on drug abuse at the state level."

The results for California and Arizona were interesting. Past month illicit drug use rates were higher in California (8.3 percent) and Arizona (8.4 percent) than in the remaining states (6.1 percent). But among youths age 12–17, the rate was lower in California (9.1 percent) and much higher in Arizona (16.8 percent) than in other states (11.9 percent).

1998 NHSDA

Two new modules administered to adults were added to the questionnaire in 1998. A "Social Environment" module was similar to the Youth Experiences module, but focused on risk and protective factors relevant for adults. Another module was a "Parenting Experiences" module. This module asked parents questions related to their child's substance use, including whether their child used drugs and whether they had talked with the child about substance use. These questions were asked only when both a 12–17-year-old and their parent were selected for interview. There was no special sample selection procedure incorporated into the household screener to identify and then sample these pairs of household members based on their relationship. In households where a youth and adult were selected, the adult interview included a question asking if they were the parent of that sampled youth. If they answered "yes," then the Parenting Experiences answer sheet was given to them to fill out. This method of sampling resulted in a subset of adult respondents that required a separate weight calculation for making nationally representative estimates using items in this module.

California and Arizona were again oversampled in 1998, for all four quarters. The final sample size was 25,500 (including 4,903 in California and 3,869 in Arizona) and the response rate was 77 percent.

Results were released in a press conference on August 18, 1999. The initial report format was nearly identical to the prior reports. But there was one noteworthy change. The title was changed from "Preliminary Results from . . ." to "Summary of Findings from . . ." It was an acknowledgement of our confidence in the quality control procedures for the data. We had had several years of error-free data. With the data checking required in the contract, the specific analyses designed to identify anomalies in the data,[17] and the OAS staff's review of tables, the consensus was that the estimates we produced for the press release each year could be considered final.

The results were summarized by Secretary Shalala:

I'm pleased to join General McCaffrey to announce the results of the 1998 National Household Drug Survey. I've said in the past that the Household Survey is a snapshot of where we are at any given moment in the fight against illicit drug use. That's true this year as well. However, it's also true that by conducting this survey over many years, we not only see where we are – we see where we're going. Two years ago I noted that we were starting to see a glimmer of hope. Today that glimmer of hope is burning bright enough that we can say: In the battle against illicit drugs – we've turned the corner. This year's survey shows that illicit drug use fell from 11.4 percent to 9.9 among young people ages 12 to 17 – a statistically significant decline, while illicit drug use among the overall population remained flat. The survey also shows that the rate of young people reporting that they tried marijuana for the first time went down – and the average age of first use went up. It's noteworthy that the survey found that the percentage of teens currently using marijuana declined from 9.4 percent in 1997 to 8.3 percent – although this was not a statistically significant decrease. All of these changes are important evidence that the perception of marijuana as a dangerous substance may be leveling off after years of decline. Sending a tough message against drugs – especially to young people – is a little like sending a message into deep space. The message goes out and then you wait a year or more to find out if it's been heard.

The 1998 NHSDA marked the end of an era. When the results were released, the 1999 survey was already in the field with a much larger, state-based sample, and computerized data collection. Nevertheless, the data already collected since the previous sample expansion in 1991 had been valuable in tracking important trends during the decade, and left researchers with an incredible data set on which to study epidemiology and policy questions concerning substance abuse. Between 1991 and 1998, nearly 200,000 respondents had been interviewed, including 70,000 youths age 12 to 17.

Notes

1 Protocol Changes 1992–93, Methodological Resource Book, 1993 NHSDA.
2 Rydell and Everingham, *Controlling Cocaine.*
3 Panel members were Dennis Nalty, Ron Simeone (Abt), Phil Leaf (Johns Hopkins), Pat Ebener (RAND), Blanche Frank (New York), Richard Spence (Texas), Steve Martin (U. Delaware), Jim Hall (Upfront), Robert Dorwart, and Len LoSciuto (Temple U.)
4 The randomization occurred at the household level, so that within a segment, both A and B questionnaires could be administered. This design maximizes the precision of comparisons between estimates from the A and B samples.
5 This was based on estimates from the Drug Abuse Warning Network that showed increases in drug-related emergency department visits.
6 Carnevale and Murphy, "Matching Rhetoric to Dollars," 306–10.
7 SAMHSA, *Preliminary Results from the 1996 NHSDA*, 98.
8 SAMHSA, "The Development and Implementation."
9 SAMHSA, *NHSDA Main Findings 1995*, Appendix E.
10 Mark Weber became SAMHSA's Director of Public Affairs in July 1996, and remained in that position until January 2012.
11 "Youth Crime: The Battle of the Reports," *US News & World Report*, August 19, 1996, 8.
12 "Skyrocketing Drug Use," *Washington Times*, August 9, 1996, A20.
13 Walters, "The White House Smoke Screen on Drugs," *Washington Times*, August 23, 1996. Walters was Deputy Director and Chief of Staff at ONDCP, and served as interim acting director after William Bennett resigned in 1990. In 1996 Walters was president of the New Citizen Project. Later he became Director of ONDCP during the George W. Bush administration.
14 Lee, *Smoke Signals*, 238–54.
15 Lee, 252
16 SAMHSA, *Preliminary Results from the 1996 NHSDA.*
17 For example, the six-month table production allowed us to do early checks on data processing and estimation. The Editing and Imputation Evaluation Report flagged any unusual changes in full-year estimates caused by editing or imputation.

8 Better Sample, Better Analysis, but Not Always

As ONDCP had intended, the sample expansion and annual surveys beginning in 1990 improved the federal government's capability to track trends and study policy issues in drug use. Combining data from 1991–93 results in about 88,000 completed interviews for analysis. This is twice the number of completed interviews in all of the National Surveys from 1974 through 1988. A few examples of studies made possible by the survey expansion are described in this chapter.

State and Metropolitan Area Estimates

The 1991–93 surveys were designed to provide estimates for the nation and six metropolitan areas. But the size of the sample made it possible to develop statistical models that could predict substance use prevalence in areas with little or no sample data. The methodology, used by several federal agencies conducting national surveys, is referred to as small area estimation (SAE). Aware of the interest from ONDCP and SAMHSA in state and metropolitan area estimates, RTI proposed a plan to create these model-based estimates with NHSDA. The developmental cost was high, but when the plan to produce quarterly drug index estimates was nixed, the funds that had been designated for it were shifted to the SAE work.

The method proposed by RTI involved "composite estimators" that would be produced for geographic areas in which there was a sample large enough to compute a reasonably precise direct estimate. It was determined that a sample size of at least 400 interviews and forty sample segments was needed for each area. In addition, at least four primary sampling units (counties or metropolitan areas) were needed for state estimates. Based on these minimum sample requirements, the method produced estimates for twenty-six states and twenty-five metropolitan areas. The estimation involved development of national prediction models of the relationship between substance use and a variety of predictors, using the entire national samples for 1991–93. Dozens of predictor variables were used, including data on local area demographics from the decennial census (such as the high school dropout rate in the census tract), and county level characteristics (such as drug treatment rates, rates of arrest for drug possession, and alcohol-related death rates). Models generated predicted prevalence rates for each state or metropolitan area. In essence, the predicted rates were combined with the direct estimates for the areas (computed in the standard way with sample responses from the area) in a weighted average. For each area, the larger the sample, the more weight was given to the direct component, relative to the model component. One way of describing this type of estimation is that it is a standard direct estimate from the sample data, but with an adjustment that smoothes the estimate by borrowing strength from the full sample to increase accuracy. If the small sample in an area results in an estimate that is far higher or lower than the norm in other areas with similar characteristics, the model component will "shrink" the estimate to be closer to what is expected, based on the values of the predictor variables for the area.

The results were published in September 1996.[1] The report generated a lot of interest, but had limited usefulness for policy purposes because it did not provide estimates for *all* states. However, the study showed the capability of the survey to produce state estimates. The work that was done to develop and carry out the methodology for this pilot study was beneficial to SAMHSA. The NHSDA team, as well as RTI's SAE team, learned a lot from this exercise.[2] It prepared us to continue the SAE work as an ongoing, integral component of the survey two years later.

Miami Success, Thanks to Andrew

It is gratifying to see the data you produce being used by national leaders and researchers. This is especially true when the people who recommended or required that certain data be collected make use of that data. But it can be disappointing and embarrassing when the data that someone had specifically asked for and was eagerly awaiting turns out to be flawed.

NIDA Director Schuster had decided in 1991 to continue the oversampling in the same six metropolitan areas, based on the recommendation of Jim Burke, Chairman of the Partnership for a Drug Free America. Burke wanted to make sure it would be possible to track drug use rates in a few metropolitan areas over time, and to study the impact of community drug prevention activities in these areas. An opportunity to do that arose in 1996. Urged by Burke, President Clinton and ONDCP Director Barry McCaffrey traveled to Miami to announce the release of the new

National Drug Control Strategy on April 29. The venue was chosen because of the apparent success of the drug abuse prevention coalition in Miami.[3] "We are here because of what Miami has done," Clinton said. "The Coalition for a Safe and Drug-Free Community has worked hard with all the rest of the people here so that drug use dropped more than 50 percent between 1991 and 1993 alone. That was the biggest drop of any metropolitan area in the country."

The use of the NHSDA data at the Miami event was troubling. For one thing, we only had data for six metropolitan areas, so we could not say whether similar or even larger declines had occurred in Houston, Philadelphia, or other metropolitan areas that the NSDUH did not oversample. Secondly, we are always uneasy about inferring causation based on the survey data. We regularly pointed out that this is a descriptive survey. It can show relationships and correlations, but it is not designed to determine causes for increases or decreases, or explain definitively why there are differences in drug use rates across geographic areas or demographic groups.

Of most concern was whether the 50 percent decline in Miami was even real. There were questions about the accuracy of the trend data from 1991 to 1993 in Miami. Unbeknownst to the president and his team, back at SAMHSA we were in the midst of an evaluation of the data. Dr. Len Saxe, a researcher who had studied the NHSDA Miami data, had contacted us prior to the president's appearance, to warn us about anomalies he had discovered in the data that were possibly related to Hurricane Andrew, which had devastated parts of Miami in late August 1992. Many previously listed households were either demolished or vacant. Saxe was aware of the upcoming mission to Miami and was hoping we could intervene in some way. We asked RTI to start an investigation, including an assessment of the effects of Hurricane Andrew on the survey estimates. We inquired with our communications office (Mark Weber) about how to inform the White House about the potential problem with the data. Weber immediately informed SAMHSA administrator Chavez and the HHS Communications Office. Despite our warnings, the Miami event went on as planned.

The RTI report on the hurricane was completed and published two months after the Miami press conference. The report documented drops in eligibility rates, response rates, and respondent cooperation following Andrew, particularly in the most heavily damaged areas. Perceived availability of drugs also declined. The analysis did not fully support any single explanation for the decline in drug use, but concluded that Andrew clearly had an effect on the trend. Although it is possible that there was truly a decrease in drug use from 1991 to 1993, the report recommended using these data cautiously, due to the hurricane's impact on the target population and data collection efforts. Conclusions should be verified with data from other sources.

Although inconclusive, the study demonstrated the value of the quarterly sample design that had just begun in 1992. This design permitted analysts to split the NHSDA sample into "before Andrew" and "after Andrew" subsamples. The hurricane occurred in late August, after third quarter data collection was nearly complete. Fourth quarter data collection did not begin until October 1, after Andrew. The ability to split each year's sample into four separate nationally representative samples would be helpful in several later studies attempting to understand the effects of specific events on substance use and mental health in the United States, especially after the state-based design began in 1999.

Several months after the pronouncements in Miami, Michael Moss, a *Wall Street Journal* reporter, wrote a story about the Miami event and the problem with the data. He interviewed SAMHSA spokesman Mark Weber, who explained what happened and shared a copy of the RTI report with him. Quoting the report, Moss' article said "There is considerable evidence to suggest that Hurricane Andrew did affect the results." The title of the article was "Miami Triumphs Over Drug Use, But Don't Try This in Your Town."

New Data for HIV-AIDS Research

The expanded sample attracted other agencies interested in collaborating with SAMHSA to collect data they needed. In the early 1990s, CDC was exploring ways to get more data on HIV-AIDS risk behaviors. In particular, there was a lack of information on sexual behaviors among various population subgroups. The initial request for sexual behavior questions came in early 1994 when Ron Wilson from NCHS, a component of CDC, contacted me. CDC was interested in adding questions to NHSDA asking about vaginal, oral, and anal sex, condom use, number of sexual partners and gender of partners, blood donation, HIV testing and HIV status. They wanted the questions included in the 1995 NHSDA and subsequent surveys. As Ron and I worked through the issues and possibilities, a formal request came in a memo from CDC Director David Satcher to acting SAMHSA Administrator Elaine Johnson. As the federal government's lead health statistical agency, NCHS conducted several large national population surveys that could have included these questions. But there were at least two reasons that NCHS preferred NHSDA. First, NHSDA already included questions on drug use with needles and needle sharing, major HIV risk behaviors.

Second, NCHS recognized the sensitivity and controversial nature of sexual behavior questions. NCHS was concerned about the effects these items might have on the other important health data that NCHS surveys capture. Respondents might object, refuse to participate, or complain to their representatives. Interviewers might refuse to administer the questions. Some of the NCHS surveys, including the National Health Interview Survey, used Census Bureau interviewers for data collection. Years earlier, requests by NIDA to add illicit drug use questions to the National Health Interview Survey had been met with resistance due to the concern that the Census Bureau would not allow their interviewers to ask for such sensitive personal information. The NHSDA, on the other hand, was designed to obtain sensitive data. Interviewers as well as design staff knew how to administer sensitive questions. Private self-administered answer sheets were used. Satcher's memo of March 17, 1994 explained:

> A set of approximately 12 questions was developed and tested in the NCHS Questionnaire Development Laboratory last year for possible inclusion in the National Health Interview Survey (NHIS). However, it became evident that these questions were not appropriate in the context of the family style interview of the NHIS. On the other hand, the NHSDA, a large, respected, ongoing national household survey that focuses on the sensitive topic of drug use, represents an ideal vehicle for these questions.

Nevertheless, we had the same concerns about the effects of adding these questions to our survey. These questions were more sensitive and controversial than anything in NHSDA. We were very careful not to allow any change in the survey that would affect drug use reporting and result in another loss of comparability across years. Tracking trends was paramount. Therefore, we agreed to add the sexual behavior module only if a field test demonstrated that the new questions would not affect other NHSDA data. SAMHSA also requested that CDC transfer funds to SAMHSA to pay for the questionnaire development, field test, and annual inclusion in the survey. While the field test stipulation would result in a delay in collecting these important new data, CDC understood our position and agreed to fund just one year of data collection, as well as the field test, and several rounds of cognitive testing to refine the question wording so we would be getting valid, reliable data.

Trepidation about adding these controversial personal questions to a major federal survey was pervasive. Before we were allowed to submit the field testing plan to OMB for approval, newly appointed SAMHSA Administrator Nelba Chavez wanted to make sure her boss agreed. Submitting a request for clearance to OMB would be a public announcement of SAMHSA's plans. The clearance process included an opportunity for the public (including advocacy groups, politicians, and news reporters) to object or otherwise comment on the plan. Nelba sent a memo to Dr. Phil Lee, Assistant Secretary for Health, asking for concurrence. In the memo, she said:

I would like your advice as to the appropriateness of proceeding with this project at this time. While I am convinced that the information would be very useful, I am concerned about the potential for controversy about the explicit nature of the questions. As well, inclusion of these questions might harm our only source of national data on substance abuse. On the other hand, I recognize the importance of collecting information about the risk factors for AIDS. This is a particular concern regarding the populations we serve. Where should we strike the balance?

The response from HHS a few days later was to proceed with the field test. I was told Dr. Lee gave verbal approval, but he did not sign his name to a document showing he had approved the sexual behavior module. Deniability therefore was a possibility.

The field test was done during the second quarter of 1995. No problems associated with administering the sexual behavior module were found, so the questions were included in the 1996 NHSDA. The answer sheet handed to respondents was innocuously titled "Personal Behaviors."

How Many People Need Treatment for a Substance Abuse Problem?

Given SAMHSA's responsibility for managing the SAPT Block Grant, estimates of how many people need and receive treatment are critically important to the agency. National estimates were needed by SAMHSA for budget and planning purposes. ONDCP needed the data to estimate the drug abuse "treatment gap," the number of people that needed but did not receive treatment. State-level estimates of alcohol and drug treatment need were also important. The law that created SAMHSA specified that states were required to explain in their applications for block grant dollars how they would allocate the funds within their state, using data on treatment need. Responding to these data requirements, OAS started exploring methods to obtain national estimates of drug abuse treatment need. Soon after SAMHSA was created, I was given the task of developing a definition and estimation method. I enlisted Joan Epstein to work with me on the project. Given the urgency, we decided to focus on data readily available from the NHSDA. We looked at previous studies[4] and explored new approaches with NHSDA. The large samples in the 1991–93 NHSDAs contained enough cases for us to test alternative definitions and assess the capability of the survey to make reliable estimates. Our approach was to define treatment need in a way that coincided with current treatment policies and practices, and account for the expected underestimation of hard-core drug use with NHSDA. After several months of analysis, reviews by others in SAMHSA, and meetings with ONDCP, a final definition was determined.[5] Treatment was needed if a person had drug dependence, used marijuana daily, used cocaine weekly, used heroin at least once, used any drug with a needle at least once, or received treatment for a drug problem in the past twelve months. Two levels of treatment need were specified. Level One represented those with a less severe problem, those who "could benefit from treatment." Level Two included those who were most in need of treatment. The survey contained all of the items necessary to classify respondents into these categories. We incorporated a ratio estimation method that adjusted the sample weights to account for undercoverage of heavy users in treatment and among arrestees.[6] Estimates for 1991 to 1996 were about nine million persons per year, including four million with Level One and five million with Level Two treatment need.

The use of this methodology continued for several years. Estimates appeared in special studies done by OAS, including the state and metropolitan area report discussed earlier in this chapter. SAMSHA and ONDCP used the estimates in developing budget planning documents each year. A disadvantage of the method was that the definition we used was particular to the survey. It was not based on a standard, commonly recognized measure of drug abuse. In 1998, an ONDCP interagency workgroup on data issues reviewed other options.[7] The workgroup favored a shift to a simpler, more widely accepted definition of treatment need: having a drug use disorder (dependence or abuse), based on the DSM-IV criteria. Joan and I met with the workgroup to discuss the feasibility of obtaining these estimates from NHSDA. The questionnaire did not include all of the items necessary to estimate whether a person had a drug use disorder, but we said we could have the survey updated by 2000. One of the concerns of using the presence of a drug disorder as an indicator of treatment need was the large population that report receiving treatment in the past twelve months but don't meet the criteria for a drug use disorder. The workgroup decided that this population of persons receiving treatment despite the absence of a drug use disorder should be included in the overall estimate of treatment need, as long as they had received "specialty" drug abuse treatment (treatment for a drug problem at a drug rehabilitation facility, at a mental health center, or as a hospital inpatient). Later analysis showed that most of this population that had no disorder in the past twelve months either had a level of problems just below the threshold for being classified with a disorder, or they had entered treatment through the criminal justice system, or their treatment began more than twelve months ago and they were possibly in post-treatment recovery.

These estimates of the number of people with drug abuse problems and therefore needing treatment are critical for policy purposes. But a policy goal of providing treatment for several million drug abusers is not practical, either politically or financially. At the workgroup meeting, ONDCP referred to "effective demand" for services, which might be defined as those who acknowledge they need treatment, or those who actually seek treatment, or perhaps those who could be convinced to seek treatment. ONDCP was envisioning a policy targeting the people who were most likely to benefit from expanded treatment availability: those who recognized their problem and would take advantage of treatment if it were available. The ultimate definition derived from the NHSDA was those who needed treatment based on their reported symptoms, but did not receive specialty treatment in the past twelve months, and also reported that they felt they needed treatment and tried to get treatment. This population of interest, and consequently the number captured by the survey sample, was small. This example illustrates that no matter how big the survey became, there were always demands for more detailed analyses and estimates of specific populations of interest that pushed the limits of the survey's capability.

When the new treatment need estimates were first available, after the 2000 survey, they quickly became established as critical indicators for tracking progress and developing federal policy. Released in a special "Report to the President" in 2001, they were used as the basis for treatment expansion initiatives such as SAMHSA's "Access to Recovery."[8] The overall drug abuse treatment need estimate was 4.7 million people. Of those, 17 percent (800,000) had received specialty treatment in the past twelve months. Of the 3.9 million who had not received treatment, only about 10 percent (380,000) felt they needed treatment. The treatment expansion focused on the 130,000 who felt they needed treatment, and tried but failed to get treatment.

Several years after the interagency work group established the definition for drug treatment need, a trio of consultants was asked to help SAMHSA decide on the official definition of need for treatment for an alcohol problem. The answer was obvious and unanimous: alcohol use disorder, or specialty treatment for an alcohol problem in the past twelve months.[9]

To Predict the Future, You Must Understand the Past

As the sample size expanded, making the data set more useful for research purposes, OAS enhanced its internal analysis capability and also took steps to make the data more accessible to outside researchers. OAS awarded a separate contract specifically to analyze the data from NHSDA and other sources.[10] The contract tasks also included the maintenance of a data archive. This was a repository where the public use microdata files from NHSDA and other substance abuse studies were made available at no charge to researchers.

OAS launched several new analytic studies that were published as SAMHSA reports and peer-reviewed journal articles. A study of youth drug use by family characteristics showed that youths in single-parent households had higher rates of drug use than youths in two-parent households.[11] A report on worker drug use by occupation and industry found high rates of drug use among construction, food preparation, and food service occupations. Police and teachers had low rates of use.[12] An analysis of college-age youths found that college students not living with parents were the heaviest drinkers, those who had not completed high school had the highest rate of cigarette use, and college students living with their parents had the lowest rate of marijuana use.[13]

The larger sample of older adults facilitated computation of drug use incidence by sex, race/ethnicity, and birth cohorts as far back as 1919. The study showed the greater use of marijuana and other illicit drugs among the baby-boom and later generations.[14] The depth of these results on patterns and trends in incidence spawned my idea for a new study that would connect the incidence results with the treatment-need data to predict future treatment need. The analysis would serve at least two purposes. It was well known in the substance abuse field that substance abuse problems and the need for treatment typically develop over a long period of time. Youths who try drugs for the first time may not experience serious addiction problems until years later. Although "Just Say No" was no longer the government's favored prevention message to youth, it seemed to me that a quantified statistical connection between trying drugs and developing a serious drug problem would be a powerful finding that could support SAMHSA's prevention approaches. Secondly, the projections of future treatment need would inform policymakers and planners of the need to prepare for the coming surge of aging baby boomers with a history of drug use. The statistical connection was made with a logistic regression model that estimated the probability that a person would need treatment, given their drug use history and demographic characteristics. Most importantly, a person who has never used drugs has a zero chance of needing treatment for a drug problem. For people who had used drugs, those who had initiated at an earlier age were more likely to need treatment than those who had initiated at a later age. The large NHSDA sample of adults, with their reports of age at first use of different substances, made it possible to compute these probabilities of needing treatment. The probabilities were applied to the sample as they aged to project treatment need as far ahead as 2020. By changing assumptions on how many youths will initiate drug use in future years, the study was able to estimate that if current marijuana initiation rate persisted, treatment need would grow by 57 percent in 2020. However, the number in 2020 would be reduced by a million persons if the future initiation rate could be lowered by 25 percent.

At SAMHSA, there often seemed to be a bias against research studies like this. It may have been a lingering attitude from the reorganization: NIH does research, SAMHSA does not. Or it may have been simply SAMHSA leaders' preference for results to be publicized as SAMHSA products, ensuring the agency and not the author gets the recognition. Some SAMHSA leaders seemed to have the belief that peer-reviewed journal articles are done primarily for personal career enhancement, instead of to help fulfill SAMHSA's mission. Administrators, center directors, and other officials rarely mentioned journal articles done by OAS. Press releases announcing these studies almost never occurred. There were many missed opportunities. Important studies could have been publicized to help advance SAMHSA's messages of prevention and treatment, but were ignored. The treatment need projection study was an exception. Published as a journal article, agency leaders recognized the relevance of the findings to SAMHSA's mission and decided to promote it.[15] A press release announcing the publication of the study was issued by SAMHSA. It was also highlighted as the "Fact of the Week" in HHS's weekly memorandum for the president on April 15, 1999. The press release included a statement by SAMHSA Administrator Nelba Chavez:

> This study reinforces the need for the President's long term, comprehensive national drug control strategy and supports the Administration's budget request to Congress to fund programs needed to close the treatment gap and provide prevention services. It is clear reducing drug abuse in America will require a long-term commitment of leadership and resources for treatment and prevention. These data can guide decisions about the levels of funding required to identify and address family, school, and mental health problems before they lead to substance abuse and the types of treatment programs that would be most beneficial.

By the time the treatment need projection paper was published, the first quarter of the 1999 NHSDA had been completed, with an even larger sample, covering every state, and using a new computerized interviewing technique. The capabilities for analysis would soon be enhanced again. But the transition to the new, improved survey would not be a smooth one. Unanticipated implementation difficulties would interfere with the best laid plans.

Notes

1 SAMHSA, *Substance Abuse in States*.
2 The lead OAS analyst on the SAE work was Doug Wright. Ralph Folsom supervised the work at RTI.
3 Attorney General Janet Reno and Treasury Secretary Robert Rubin were both from Miami, so they joined Clinton and McCaffrey at the event.
4 Gerstein and Harwood, *Treating Drug Problems*.
5 SAMHSA, "Analyses of Substance Abuse," 113–45.
6 Wright, Gfroerer, and Epstein, "Ratio Estimation," 401–16.
7 The workgroup was chaired by Terry Zobeck of ONDCP. Agencies participating included ONDCP, SAMHSA, NIDA, NIAAA, National Institute on Justice, and Bureau of Justice Statistics.
8 US Department of Health and Human Services, *Closing the Treatment Gap*.
9 The advisors were William McAuliffe (Harvard Medical School), Jane Maxwell (University of Texas), and Robert Wilson (University of Delaware). SAMHSA attendees at the June 3, 2002 meeting were Steve Wing, Joan Epstein, and Joe Gfroerer.
10 The contract was recompeted every few years, with RTI and NORC alternately winning the contracts.
11 Johnson, Hoffman, and Gerstein, *The Relationship Between*.
12 Hoffman, Brittingham, and Larison, *Drug Use Among US Workers*.
13 Gfroerer, Greenblatt, Wright, "Substance Use in the College-Age," 62–5.
14 Johnson, Gerstein, Ghadialy, Choy, and Gfroerer, "*Trends in the Incidence*."
15 Gfroerer and Epstein, "Marijuana Initiates," 229–37.

Wrap-Up for Chapters 6, 7, and 8

Major changes to the NSDUH design occurred during the 1990s, driven by several factors. The crack epidemic and cocaine overdoses prompted legislation that led to sample size increases and annual data collection. Passage of marijuana legalization bills in two states prompted HHS to increase the NHSDA sample sizes in those states, and push for a redesign of the sample to produce estimates for all states. Other more subtle changes to the survey were developed by the NHSDA team and contractor. A twelve-month field period with quarterly samples improved the efficiency of data collection and processing. The program of methodological development and testing led to an improved questionnaire. Revised data processing procedures resulted in better data quality and more reliable estimates. A more predictable schedule for data processing and analysis, including a full report of findings and methods completed prior to data release each year, demonstrated transparency and helped to depoliticize the reporting of results.

All of these improvements, coupled with the placement of the survey in SAMHSA's newly formed Office of Applied Studies (OAS), contributed to an enhanced reputation of the survey and recognition of its statistical integrity. HHS Secretary Donna Shalala was supportive of the survey, understood its strengths and limitations, and was instrumental in its continuing expansion and improvement. Other organizations recognized the power of the survey as a platform for obtaining data they needed, providing funds to add new modules of questions. Although the survey had only a very small staff when it was moved to SAMSHA, the agency recognized the importance of having a strong in-house staff to manage the project. OAS was allowed to recruit and build a team of survey methodologists and analysts.

The focus of the survey began to shift towards areas of primary concern to SAMHSA and its three centers. Tracking of treatment need and treatment became a high priority, given the CSAT responsibility for managing the large block grants to states. Although CMHS had little interest in collaborating with OAS on the collection of mental health data, SAMSHA administrators and OAS directors wanted the survey to collect data on mental health. Mental health modules for youths and adults were added in 1994, but they were removed from the questionnaire after 1997, at the request of CMHS. The high-level push for the survey to cover mental health persisted, and modules were included in 2000 and subsequent years. SAMHSA administrators also conveyed that prevention-related topics should be addressed by the survey. OAS collaborated with CSAP to develop a module on risk and protective factors for adolescent substance use, introduced in the 1997 survey.

SAMHSA's creation also began an era of cautious cooperation between ONDCP and OAS. The relationship was symbiotic. ONDCP depended on OAS to provide timely access to the data and quick turnaround of special analyses. The OAS wanted the data to be used by ONDCP, which would solidify the recognition of the value of the data. OAS also depended on support from ONDCP when budgets were developed and proposals to cut NSDUH funding emerged. To circumvent bureaucratic delays (such as during periods in which HHS or SAMHSA leaders required all communications between ONDCP and the data office to flow through them) or when there were policy or budget disagreements between ONDCP and HHS, a back-channel line of communication was at times employed, to allow ONDCP to get data or draft reports from OAS quickly and secretly.

Increases in the sample size, analysis budget, and staffing led to greater use of the data for policy and research purposes. OAS began producing a series of analytic reports through a separate contract with the National Opinion Research Center. Studies focusing on substance use among women, racial/ethnic subgroups, college students, and youths were published. Important research issues that were addressed included the association between mental health and substance use; the relationships between socioeconomic status and family structure with substance abuse; risk and protective factors for drug use among youth; children with substance-abusing parents; substance use during pregnancy; and historical patterns of drug use by birth cohorts. Policymakers increasingly relied on the NHSDA data as the survey improved. ONDCP analysts frequently asked OAS to produce special tabulations on topics they were investigating, and NHSDA data were cited in the annual National Drug Control Strategy. ONDCP analysts relied on and trusted the statistical expertise of the NHSDA team in OAS, and sometimes consulted with them when they were developing analytic plans. The survey became a primary data source used by ONDCP and SAMHSA in setting policies regarding federal support for drug abuse treatment. Estimates of the number of people needing and receiving treatment were generated from the survey. NHSDA data were also instrumental in identifying and describing the increases in drug use among adolescents during the 1990s. The data helped SAMHSA and HHS develop policies to prevent further increases.

Use of the data was also expanded by OAS's creation of a data archive. The archive made NHSDA and other data sets easily accessible to outside researchers, with rigorous controls to protect the confidentiality of survey respondents.

Despite the OAS efforts to enhance the statistical integrity of the survey and to produce unbiased reports that provided guidance for interpreting the findings, politicians and their supporters continued to "spin" the data for political purposes. When the Republican Bush administration reported 1990 NHSDA results showing declines in drug use, administration officials took credit for implementing policies that caused the decline. Democratic leaders responded by conducting their own studies to show the NHSDA numbers were wrong. Five years later, during the Democratic Clinton administration, the survey showed increases in drug use among youth. Republicans cited the data as evidence that Clinton's policies were failing. The Clinton administration countered by citing NHSDA and MTF data indicating that the increases began during the Bush administration. Then, just a few months before the 1996 election, Clinton highlighted the decreases in drug use in Miami (despite the data problem due to Hurricane Andrew), and attributed them to prevention efforts supported by the administration. Of course, these statements were not the result of a collaboration or even a consultation with the NHSDA team of statisticians. OAS made efforts to provide guidance in terms of the precision of estimates, statistical significance of differences, and caveats associated with the survey data. This information was documented in the SAMHSA reports made available with each year's data release and subsequent reports. And the survey team was always willing to discuss the results with data users that contacted OAS. But the survey team was careful not to attribute causation.

NSDUH and most other large surveys are designed to provide mainly descriptive data, not explanatory data. The survey tells us with near certainty that the proportion of youths in the country using marijuana was greater in 1995 than in 1992, but it does not tell us why. The survey results cannot tell us that an increase or a decrease in use was due to a particular government policy or program. There are analytic techniques and study designs that can legitimately provide evidence of causation, but these are not typically incorporated into large population surveys like NSDUH. The politicians and their advocates who report survey statistics are sometimes careful to make sure the data are reliable, in terms of the accuracy of the estimates and significance of differences. Statisticians should provide support for these data users. But when politicians use the data to claim their own success or the opposition's failure based on unfounded interpretations and assumed causes, statisticians and others responsible for the survey must take the side of statistical integrity, and not support these interpretations. That does not necessarily mean actively opposing or publicly denouncing the incorrect uses of the data. But it does mean that if asked, statisticians should not agree with them. An appropriate, safe response which I have used in these situations is to say that the survey is not designed to determine causes. Often, there is evidence from the survey results that suggest certain factors may be contributing to the increase or decrease, and those can be mentioned, with appropriate caveats. For example, NSDUH and MTF reports have regularly included trends in perceived risk of harm in using drugs, along with the trends in actual rates of drug use. The striking correlation between these two trends is an important finding that *suggests* that changes in use may be affected by youths' beliefs about the health effects of use of the drug. But we would not report these results (or recommend that the drug czar or other agency directors report) as definitively as "Rates of drug use are increasing among youths because fewer youths believe they are dangerous." Instead, reports introduce these data by saying "One factor that can influence whether individuals will use tobacco, alcohol, or illicit drugs is the extent to which they believe these substances might cause them harm."[1]

Note

1 Lipari, Kroutil, and Pemberton, *Risk and Protective Factors.*

9 A Perfect Redesign Storm

During the 1990s, the survey became more widely known and respected as an important data source on substance use in the United States. With a larger sample, continuous data collection, and an improved questionnaire, demands for analysis increased. But the desire to make more improvements to the survey continued. Survey methodologists recommended modernizing the survey to a computerized format. ONDCP and HHS wanted the survey to provide estimates for every state. Congress and the White House wanted the survey to determine multimillion dollar fines for tobacco manufacturers marketing their products to children. These distinct demands on the survey converged in a comprehensive redesign of the NHSDA in 1999.

Computer-Assisted Interviewing

The questionnaire had been redesigned in 1994, using the results of methodological research studies completed during 1989–92. Those studies showed that the methods of collecting data used in the survey since 1971 were generally good. Herb Abelson's self-administered answer sheets, limited use of skip patterns, and other procedures worked well. The only changes made in 1994 were to refine the wording of some questions, reorganize the questionnaire, and shift the tobacco questions from interviewer-administered to the superior self-administration technique. Data editing procedures were revised. At the time, another change to the survey was discussed but not implemented. RTI recommended converting the survey to Computer Assisted Interviewing (CAI). For interviewer-administered portions of the interview, Computer Assisted Personal Interviewing (CAPI) would be used. Instead of reading and filling in answers in a questionnaire booklet, interviewers would read questions from a laptop computer screen and enter answers using the keyboard. The self-administered part of the NSDUH would be done using a relatively new technology called Audio Computer Assisted Self Interviewing (ACASI) that RTI was at the forefront of developing.

Instead of marking answers to sensitive drug use questions on paper answer sheets, respondents would read questions on the computer screen and enter their answers using the keyboard. Respondents could hear recorded audio of each question using headphones, as each question appeared on the laptop screen. The advantages of CAI over paper and pencil interviewing (PAPI) were clear. Keypunching of responses from the paper instrument would no longer be needed, because all of the data would be captured electronically during the interview. Persons selected for the sample who could not read well enough to complete the self-administered answer sheets would be able to participate by relying on the voice recordings. For respondents, ACASI offered more privacy than the paper answer sheets. Observers of the ACASI part of the interview, as long as they cannot view the computer screen, could not ascertain which questions the respondent was answering at any moment, and the answers entered by the respondent were gone from the screen as soon as they moved to the next question. The interview skip patterns could be much more complex, since the decision on which question needed to be answered next would be directed by a computer program that would automatically bring onto the screen the desired question, based on earlier responses that are stored and retrieved during the interview. With the enhanced privacy, skip patterns could be inserted into the substance use modules without potentially revealing to observers whether the respondent had reported they had used drugs. The CAI program also gives the capability to resolve reporting inconsistencies as they occur, with a follow-up question to the respondent, instead of months later during the editing phase of data processing.

Despite RTI's enthusiasm for ACASI, we decided not to introduce it in the 1994 redesign. There were too many unknowns. No survey as large as NHSDA had used ACASI. And of the few small studies that had used it, none had an ACASI component as lengthy as the self-administered portion of NHSDA. Research on ACASI was sparse. To convert the NHSDA from paper to CAI, years of development and testing would be needed, and a careful plan for implementation. The importance of tracking drug use over time meant we would probably need a simultaneous, large supplemental sample of PAPI interviews, to provide a bridge between old and new estimates, because we expected the change in data collection mode would affect how respondents would answer the questions.

In 1995 we decided to proceed with research and development of a CAI version of the survey. Luckily, OAS was able to hire one of the top questionnaire design experts in HHS, Peggy Barker, to lead our efforts. She had worked at the National Center for Health Statistics for more than fifteen years, and had recently spearheaded the conversion of the National Health Interview Survey from paper to computer. A long-term plan for NHSDA was developed. It included two field tests and cognitive testing. Unless major problems with the new methodology emerged, we expected to introduce the CAI with ACASI in the 1999 survey.[1]

For the first field test, 400 respondents were interviewed in 1996 using two different CAI versions of the NHSDA interview. One version had the same questions as the 1996 PAPI survey, with no skip patterns in the self-administered portion of the interview (ACASI). The other version was as similar as possible except skip patterns were built into the ACASI portion. This first test demonstrated that the CAI approach was feasible. Both versions worked well. Respondents were willing to complete thirty minutes of ACASI questions, and skip patterns reduced the interview time by ten minutes. There was no evidence the new procedures would cause a lower response rate or diminish data quality. Although about half of respondents had never used a computer before their interview, only 3 percent complained about the difficulty of doing so.

To help guide the redesign planning, we convened a group of consultants with expertise in survey design, substance abuse and mental health epidemiology, and policy.[2] Topics covered during the meeting were state and local estimation, questionnaire content, population coverage, measuring household and neighborhood effects, and field procedures. Most of the consultants were supportive of using modeling to make state estimates, and not increasing the sample too drastically. An expansion of the target population to children under age twelve was discouraged because of the low rates of drug use and difficulties with comprehension that would impact data quality. There was general agreement that ACASI would work on NSDUH and improve the quality of the data, but that we should be prepared to face many unique operational issues. These included whether separate electronic devices would be needed for household screening and interviewing; what would the interviewer do to occupy her/his time while the respondent is completing the ACASI portion of the interview; what type of voice should be used for the ACASI recordings, and how would it affect reporting. There was a suggestion to phase in the CAI over a two- or three-year period, rather than all in one year.

The CAI instrument underwent extensive refinement through cognitive laboratory testing, to prepare an updated questionnaire for a larger field test in the fourth quarter of 1997. The field test administered different versions of the CAI instrument to 1,982 respondents. The results helped determine the best structure for the skip instructions within the drug use modules, the use of data quality checks within the

ACASI sections, and whether respondents should be given multiple opportunities to report thirty-day and twelve-month use. A new handheld computer for the household screener was also tested. Based on the results of the two field tests, we decided to proceed with the conversion of the survey from PAPI to CAI beginning in 1999.

Estimates for Every State

Separate from the work on developing the computerized interview, discussion continued on the HHS Secretary's desire to expand the survey sample to provide state estimates. A primary impetus for expanding the survey sample size was the alarming increase in youth drug use during the 1990s. A larger sample would allow more detailed analysis of youth drug use to better understand the problem and more effectively target prevention efforts. State data would make it possible to study the association between youth drug use and changes in states' drug laws. SAMHSA had expanded the samples in California and Arizona for the 1997 and 1998 surveys because of the legalization of medical use of marijuana in those states. Other states would undoubtedly be passing medical marijuana laws. ONDCP was also very interested in expanding the NSDUH to obtain state estimates to help monitor the use and impact of the prevention and treatment funds given to states through the SAPT block grant. ONDCP was not satisfied with the data from CSAT's State Systems Development Program (SSDP), which supported states' efforts to conduct telephone surveys of their populations to assess treatment need. The quality of the SSDP surveys was mixed, and it was not possible to compare results across states because of the different methodologies used. Terry Zobeck, ONDCP's Research Branch Chief within the Office of Planning, Budget, Research & Evaluation, met with numerous state data representatives to explain the value of the NSDUH expansion, relative to SSDP, and gain their support. A key benefit he conveyed to them was the use of consistent methodology across all states, resulting in valid comparisons between states and with the overall US estimates.[3]

The NHSDA team was asked to propose options for revising NSDUH in order to provide estimates for each state. Doug Wright, mathematical statistician in my branch, supervised these efforts. Working with RTI's sampling statisticians, he explored alternatives for state estimation. The least expensive approach, at a cost of $10 to $15 million per year, would involve a slight expansion of the national sample to about 30,000, and the use of statistical modeling, or small area estimation, to make estimates for states. The most expensive was to expand the sample enough in every state to produce direct estimates, without resorting to the use of any small area estimation methods. We estimated this would cost $100 million, with more than 200,000 persons interviewed per year. A compromise approach, which we favored, would expand the sample to about 70,000 per year, at an estimated cost of $28 million. Large enough samples for annual direct estimates would be selected in eight large states. In other states, small representative samples would be selected to support indirect estimates using "composite" estimation, similar to the method used for the state and metropolitan area estimates from the 1991–93 surveys.

HHS convened a federal workgroup "to advise and assist SAMHSA in addressing the Secretary's initiative to develop approaches to state level substance abuse data." The group consisted of representatives from several HHS agencies, the Bureau of Labor Statistics, and the Census Bureau. The group met several times, and concluded that the compromise approach "is quite sound and compares favorably with the work of these other statistical agencies."[4] Statistical consultant Dr. Wesley Schaible, former NCHS statistician and Associate Commissioner of the Bureau of Labor Statistics, prepared the final report on the workgroup's review. In his report, he pointed out the advantage of our proposed plan to have a representative sample within each state, saying "The sample allocation, made with State estimates in mind, could help the proposed SAMHSA system be among the best indirect systems in the Federal government." But he also urged caution in interpreting results from the first year or two, until relationships and trends emerge as data accumulate over time.

Support for the approach was clear when Don Goldstone and I briefed Secretary Shalala and other key HHS officials on March 3, 1997.[5] Shalala wanted the expanded sample to begin in 1998, but Congress had not approved the funding yet. There were concerns about how the data would be used. Finally the funding for the state-based NHSDA was approved in November 1997, but "with the understanding that it must be used by the agency to improve the provision of treatment and prevention services in states with high incidence of substance abuse." Furthermore, SAMHSA was directed to distribute the results of the survey to each state, and "all states shall analyze their relative performance in preventing substance abuse as a component of the substance abuse block grant application."[6] Congress also directed SAMHSA to report to the House Appropriations Committee regarding its plans for using the NHSDA data to meet these requirements. SAMHSA submitted the six-page report in February 1998.[7]

With Congressional approval, the sample expansion was set for 1999. Fearful the sample change might adversely affect the CAI implementation, we decided, at least at first, to delay the CAI implementation until 2000, and phase it in over a two-year period. The sample would include CAI and PAPI subsamples in both years.

Tobacco Module

Late in 1997, another unrelated HHS initiative was smoldering that would soon enter the NHSDA realm, creating the perfect redesign storm with CAI conversion and the state sample design. President Clinton wanted to take action to reduce tobacco use among teens. He asked Congress to pass legislation that would set ambitious targets to reduce teen smoking, and to impose severe financial penalties on tobacco companies if the targets were not met. The penalties would be based on each company's share of the underage cigarette market and the percentage of youths currently smoking cigarettes. Congress began to take action, drafting the legislation that the president was asking for. Co-sponsored by Republican Senator John McCain and Democratic Senator Ernest Hollings, it was referred to as the McCain bill.

This legislation would require a new surveillance system that tracked cigarette use by teens according to brands smoked. In October 1997, HHS formed a workgroup to devise a plan to meet this data need. The workgroup members were survey statisticians, tobacco experts, other key people from various HHS agencies, and the Census Bureau. Meetings were chaired by the office of the Assistant Secretary for Planning and Evaluation (ASPE), but much of the discussion was led by staff from CDC's Office on Smoking and Health (OSH). By March of 1998, the workgroup had met several times and determined a new survey would be conducted. A sample of 20,000 youths age 12–17 would be extracted from the sample of households that had participated in the National Health Interview Survey. These households would be recontacted and youths would be asked to participate in the tobacco survey. Data collection would be carried out using ACASI to maximize accurate reporting by youths in the household, and the survey would start by July 1, 1998. The cost of the survey was estimated at $28 million per year, for national estimates.

Senior HHS leaders were briefed on the plan in early May 1998. Background materials for the briefing said that there was no existing survey adequate to serve as the basis of the surcharge system. The draft McCain bill proposed using MTF, but "MTF is inadequate for a number of reasons. A new, stand-alone survey is required." But SAMHSA had a better idea. Why not use the NHSDA? The sample of youths would be 22,500 each year starting in 1999. State estimates would be a bonus. Although SAMHSA had been planning to introduce ACASI starting in 2000, we offered to include a brief ACASI instrument in 1999 for the tobacco module. A few questions would have to be added to determine usual brands used. Using NHSDA would save HHS many millions of dollars. By June 3, ASPE recommended the SAMHSA plan to the Deputy Secretary, and he approved. The McCain bill was revised, naming the NHSDA as the data source for tracking youth tobacco use and determining tobacco industry penalties. On June 22, President Clinton announced the survey, five days after Republicans in the Senate, influenced by tobacco industry lobbyists, had voted to block the McCain bill. Excerpts from Clinton's statement:

> Today we're going to be talking about health concerns of American families. Of course, one of the biggest health concerns is youth smoking, something we've been discussing a lot around here lately. We all know that 3,000 young people start smoking every day,[8] and that 1,000 will die earlier because of it, even though it's illegal in every state to sell cigarettes to young people.
>
> A majority of the Senate now stands ready to join us, but last week the Republican leadership placed partisan politics and tobacco companies above our families. Their vote was not just pro-tobacco lobby, it was anti-family. The bipartisan bill they blocked would not only protect families from tobacco advertising aimed at children; it would protect children from drugs, give low and middle income families a tax cut by redressing the marriage penalty, and make substantial new investments in medical research, especially in cancer research.
>
> While we wait for Congress to heed the call of America's families, I'm instructing the Department of Health and Human Services to produce the first-ever annual survey on the brands of cigarettes teenagers smoke, and which companies are most responsible for the problem.
>
> Once this information becomes public, companies will then no longer be able to evade accountability, and neither will Congress. From now on the new data will help to hold tobacco companies accountable for targeting children.
>
> Again, I urge Congress to pass bipartisan, comprehensive legislation rather than a watered-down bill written by the tobacco lobby.

The next day, the Federal Register included official public notification of the tobacco module to be included in the 1999 NHSDA.[9] Although the fate of the McCain bill was in doubt, the ASPE workgroup continued to have regular meetings, with the focus shifted to developing the NHSDA tobacco module. RTI survey methodologists and questionnaire designers were invited to the meetings to participate in technical discussions about the survey design and the wording of questions. Ultimately, the McCain Bill never became law, but the new tobacco module remained in NSDUH. Data from the module were useful in tobacco research studies.

The Final Implementation Plan

During that same month, the NHSDA team held one of our regular management meetings with RTI. These meetings were held twice a year, usually at SAMHSA headquarters. The meetings provided an opportunity for project leaders to be updated on shifts in priorities and to identify emerging problems that needed attention. Usually the discussions were routine, consisting of a status report on sampling, data collection, analysis, budget, and other areas. But the June 1998 meeting was different. A big decision leading to a change in direction occurred. A goal of the meeting was to discuss and plan the details of the implementation of the state sample design, the CAI instrument, and the expanded tobacco module. However, as the discussion progressed it became clear that a series of uncoordinated decisions had led to a convoluted implementation plan: a huge expansion of the sample in 1999, with the old PAPI questionnaire, but with a short tobacco module in ACASI; a conversion of the survey to CAI beginning in 2000, with part of the sample continuing to use the old PAPI instrument with the ACASI tobacco module; full implementation of the CAI instrument in 2001. Concerns about this plan included the need to quickly purchase hundreds of laptop computers, the reactions of respondents to the dual use of ACASI for tobacco questions and then paper answer sheets for all other substances, the difficulty interviewers would have with this procedure, the delay until 2001 for a full CAI implementation, and the impact on trend measurement of questionnaire changes in three consecutive years. We decided that it would be better to simply make the changes all at once, in 1999. That approach would provide comparable youth tobacco data across multiple years sooner, and would result in improved data for illicit drugs two years earlier. The only downside was that it would cut short the time for updating some of the ACASI modules, especially the prescription drug modules. But the biggest question was whether we were confident that a fully tested CAI instrument covering the entire NHSDA interview could be ready by January. RTI's answer to that question was a qualified yes. It would be a huge effort. Equipment had to be acquired, a large dual sample (PAPI and CAI) had to be selected, and more staff had to be hired. The ACASI program needed to be developed, tested, and ready for use in a "training the trainers" session in early December 1998. A new case management system that tracks the status of sampled cases needed to be developed. A final field test, planned for August 1998, would need to be redesigned to test a new CAI instrument, with *all* answer sheets converted to ACASI.

How Many Federal Statisticians Does it Take?

Getting approval to hire enough staff to satisfactorily oversee the survey was an ongoing challenge. When the 1999 redesign was approved, the project was managed by my branch of about seven people. And NHSDA was not our only responsibility. Of course, much of the NHSDA work was done by the contractor. But no matter how capable contractors are, they have to be closely monitored and directed on a multifaceted project like this. SAMHSA, ONDCP, and others that rely on it expect timely, accurate data, valid trends, and responsiveness to changing requirements. Insufficient statistical staff in the agency managing a survey contract costs taxpayers money. Contract managers must continually monitor the details of many aspects of a large survey, because seemingly minor problems can quickly become major problems when hundreds of interviewers are contacting thousands of households each month. For example, if interviewers make a small change in protocol to simplify their data collection responsibilities, it could cause a small increase in the cost of completing an interview. A ten dollar increase per interview may not seem significant, but after 70,000 interviews the cost is $700,000. After three years, it's $2 million. A slight change in the wording of a question or in an editing rule could unexpectedly cause a small change in estimates – a major problem in a survey so big that a 1 percent increase from one year to the next in marijuana use is statistically significant. And if the questionnaire construction or data processing is not closely monitored, these problems may not be noticed for months, and a fix would need to be done – costing tens of thousands of dollars, and possibly a loss of trust in the survey by key users.

When it became clear the survey would soon be expanding, OAS Director Donald Goldstone elevated the NHSDA team from branch to division, creating the Division of Population Surveys, with me as director. He demanded that SAMHSA allocate more staff (FTEs, full time equivalents) to the project. SAMHSA's administrative leadership resisted, possibly because Don's combative style sometimes irritated and alienated people. But he was right. He understood the complexity and potential for cost and technical problems inherent in survey operations. However, SAMHSA was not a survey research organization, so agency leaders and personnel office staff were not always receptive to these concerns. Under these circumstances, it's important for survey staff to communicate effectively with personnel and other administrative leaders. An example of one of Don's pleas for more staff, in the context of his proposed plan for reorganizing OAS (forty-two FTEs), and being told that one of his FTEs would be given to another office to handle "data coordination":

I keep telling people we cannot carry out our responsibilities with either the staff we have or a few additions. We have a total of 3.5 FTEs working for Gfroerer with respect to the substantive aspects of the expanded household survey. I will not take responsibility for a potential disaster. We cannot make it on the 39 FTEs we have identified. Raise the request to 42 by adding an additional 3 FTEs to the Division of Population Surveys.

Dedicating one full FTE to the problem of coordinating data activities is absurd given the present commitment of a total of 3.5 FTEs to a $200 million contract activity which gets national attention. Given the size and attention this activity receives, I want things to be perfect when the Inspector General arrives for a review. That will not be possible with the number of staff now available here.

The pugnacious Dr. Goldstone did not get everything he asked for. But over the next several years, we were able to increase the number of staff in the Division of Population Surveys. By 2002 the division had about ten staff, fully devoted to the survey. Almost all were statisticians, but with different areas of expertise. They each had specific roles and responsibilities in contributing to the oversight of the contract and other activities associated with the project. Mathematical statistician Art Hughes became the new NHSDA project officer in mid-2000, providing strong oversight on all statistical aspects of the survey such as sampling, weighting, and SAE modeling. Survey methodologists were recruited to form an instrumentation team led by Peggy Barker that focused on the questionnaire and data collection. Statistical analysts wrote reports and directed the contractor's analyses. Statisticians were assigned to oversee work associated with building the data files, including editing, imputation, variable creation, and documentation. Construction of public use data files required expertise in confidentiality and disclosure limitation, balancing the utility of the data files against the need to prevent any respondent's identity from being disclosed. Managing a large survey with a staff of only ten was difficult, but it facilitated interaction between the staff with different roles. That was a benefit. Larger survey organizations may have separate branches or teams for sampling, questionnaire design, data collection, and report writing. This separation can suppress valuable interaction across the project.

In addition to adding staff to manage the expanded survey, OAS Director Goldstone decided the survey needed a permanent panel of outside consultants. This would be a group of experts we could convene as needed to get advice on current NHSDA issues. It was set up as a contract task, so officially the group was helping the contractor, but OAS chose the participants, determined the agendas, and ran the meetings. We called it the NHSDA Expert Consultant Panel. The members included experts in different areas relevant for the NSHDA, such as survey design and estimation, substance abuse and mental health epidemiology, prevention research, and drug abuse policy. The first meeting was held October 6, 1999.[10] The panel met eight times between 1999 and 2008, helping OAS make decisions on the survey design, editing and imputation, data file dissemination, questionnaire content, and other NHSDA concerns.[11] Over time, the group became knowledgeable of the survey details, making them extremely effective in giving SAMHSA relevant, practical recommendations.

Dr. Goldstone understood the survey's value and technical aspects, and recognized the importance of adequate staffing and expert advisors. He enjoyed being involved in the project. When he retired in February 2004, he spent his last day of federal service chairing an all-day NHSDA Expert Consultant Panel meeting.

The New Sample Design

The sampling plan for the 1999 NHSDA was unique. A typical multistage national household survey employs a first stage selection of large geographic areas consisting of counties or groups of counties. The nation may be divided into about 2,000 of these primary sampling units (PSUs), and a sample of a few hundred PSUs is selected at the first stage, based on their size (population). This approach provides efficiencies in staffing and data collection, because resources can be targeted at these few hundred chosen areas, controlling travel costs. But while the sampling of large clusters at the first stage is preferred for national estimates, it fails when estimates for all fifty states are desired. A typical national sample of a few hundred PSUs will contain too few PSUs in sparsely populated states like Alaska and Wyoming. In fact, in each of the NHSDA national samples prior to 1999 (which had fewer than 150 PSUs), there were several states that had no PSUs in the sample. To make estimates by state using the composite estimation method, it is preferable to have adequate representative samples of the population in every state. To achieve this for the 1999 and subsequent NHSDAs, the country was divided into 900 geographic areas. In the eight largest states, forty-eight areas of approximately equal population size were created. In the remaining forty-two states and the District of Columbia, twelve equally sized areas were defined. There would be one field interviewer (FI) assigned to cover each of the 900 areas. The areas were named FI regions. They ranged in size from 1.7 square miles (a section of the District of Columbia) to 359,000 square miles (northern portion of Alaska). In sampling terminology, the FI regions were strata, not clusters. In each FI region there would be samples of smaller areas (segments) drawn and assigned to the FI. Thus, the first stage of sampling in the multistage design was the selection of segments, small geographic areas of about 150 households, within each FI region. Eight segments per year (two per quarter) were selected in each FI region, for a total of 7,200 first stage areas. All of the selected segments were counted and listed, which means field staff traveled to these areas and constructed a list of addresses, using strict rules for starting points and the paths to follow in creating the list. Statisticians selected samples of addresses from these lists, and the addresses were assigned to interviewers for contact during the calendar quarter. As in previous NHSDA sampling plans, either no, one, or two persons were selected for an interview within each of the sample addresses. An average of about ten respondents per year would be interviewed in each segment. The resulting target sample was 67,500 respondents, with 3,600 respondents in the eight largest states and 900 respondents in the other forty-two states and the District of Columbia. The rationale for the way the sample was allocated across the states was to balance the need for both state and national estimates. Equal sample sizes across all states would result in an inefficient national design – a large design effect and relatively large sampling error. National estimates are more accurate when the sample size per state is proportional to the populations in each of the states. Since about half of the US population lived in the eight largest states, 50 percent of the sample was allocated to those states. There was no need to oversample blacks or Hispanics, since the sample would be large enough to provide reliable estimates of substance use rates for these and other minority populations. Each state's sample would be evenly spread across three age groups: 12–17, 18–25, and 26 and older.[12]

This sampling plan would remain virtually intact until 2014. However, after a few years the term "FI region" was replaced with "state sampling region (SSR)" because interviewers are often assigned to help in several regions, and the number of interviewers needed to efficiently complete all data collection does not necessarily match the number of sampling regions. In 2005, small adjustments were made to SSR boundaries, to account for updated data from the 2000 Decennial Census. Also in 2005, an extra step was inserted in the sampling. SSRs were divided into census tracts, and eight census tracts were selected in each SSR. Then one segment was selected in each sample tract. The sample design is diagrammed in Figure 9.1.

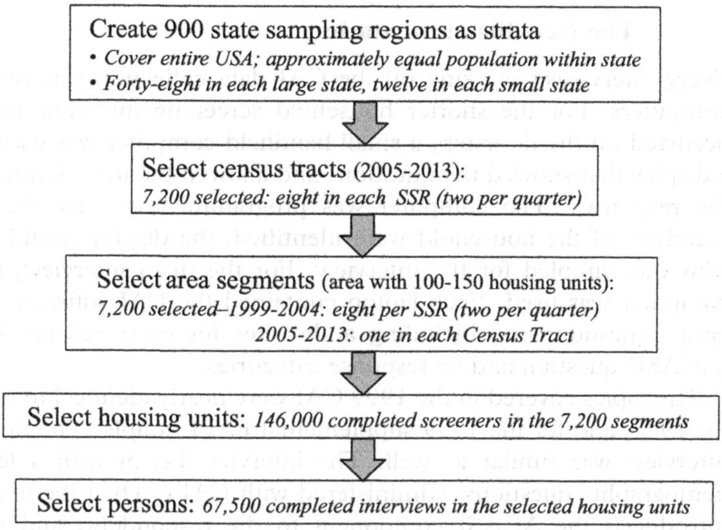
Create 900 state sampling regions as strata
• Cover entire USA; approximately equal population within state
• Forty-eight in each large state, twelve in each small state
Select census tracts (2005-2013):
7,200 selected: eight in each SSR (two per quarter)
Select area segments (area with 100-150 housing units):
7,200 selected–1999-2004: eight per SSR (two per quarter)
2005-2013: one in each Census Tract
Select housing units: 146,000 completed screeners in the 7,200 segments
Select persons: 67,500 completed interviews in the selected housing units

Figure 9.1 Multistage sample for 1999–2013 surveys.

The description of the sample in Figure 9.1 focuses on a single year. Actually, samples of segments were drawn for multiple years. For example, multi-year samples were drawn for 1999–2001, 2002–4, 2005–9, and 2010–13. Within each of these time periods, there was also an intentional overlap of samples for adjacent years. In year one of a multi-year sample, half of the selected segments (3,600) were used again in the year two sample. A new half-sample of 3,600 segments was selected for year two, and remained in the sample for year three. However, a new sample of *addresses* is selected each year. The overlapping samples of segments helps cut costs, because counting and listing does not have to be repeated in the second year that a segment is in the sample. It also complicates analysis a bit, due to the positive statistical correlation created between estimates from adjacent years. Although the impact is not large, this correlation actually improves accuracy for detecting differences in estimates from one year to the next.

The 1999 NHSDA also included a supplemental national sample with the 1998 PAPI instrumentation. This sample was originally designed to yield 20,000 completed interviews, but during 1999 it was reduced to 15,000 due to budget limitations and data collection difficulties. For this component of the 1999 survey, the 900 FI regions were used as first-stage sample clusters. A national sample of 250 FI regions was selected.

The New Instrumentation

Every interviewer working on the CAI data collection was issued two computers. For the shorter household screening interview that often occurred on the doorstep, a small handheld computer was used. It had a display that showed the questions and allowed the interviewer to enter the responses. The computer was programmed so once the eligible members of the household were identified, the display would indicate who was sampled for the interview. For the main interview, a laptop computer was used. Each laptop contained the CAI software with the entire questionnaire, including audio files for every self-administered (ACASI) question and its response categories.

The topics covered in the 1999 CAI were nearly identical to the topics for 1998, and for the 1999 supplemental PAPI sample. The flow of the interview was similar as well. The interview began with a few basic demographic questions, administered with CAPI. Then the interviewer introduced the ACASI component to the respondent, and gave the respondent a short tutorial on how to move through the questionnaire and how to enter responses. The respondent was instructed to tell the interviewer when they reached the end of the self-administered questions, at which point the final CAPI portion of the interview began, covering additional demographic questions, ending with income questions.

At the start of the ACASI portion of the interview, respondents were given headphones to listen to questions and response options. This helped poor readers and visually impaired respondents to complete the survey. The entire CAI interview, including text and voice recordings, was available in either English or Spanish. The respondent could choose which language they preferred, just as with the PAPI questionnaire in prior years.

The ACASI methodology made it possible to obtain more detailed, specific information about respondents' substance use. Most of these enhancements were possible because of the complex skip instructions that can be programmed in the CAI instrument. Examples are listed below:

- If a respondent indicated on an initial question that they had never used a particular drug, the program would skip all of the questions about age at first use, recency, and frequency of use of that drug. Later in the ACASI portion, the program would skip questions asking about symptoms of a substance use disorder for that drug.

- The ACASI portion of the interview included the expanded tobacco module developed for the McCain bill, with listings of dozens of cigarette, smokeless tobacco, and cigar brands for respondents to select from.
- For each substance, new questions were added to get more detailed information on recent first-time use of substances. If a respondent reported that their first-time use of a substance occurred recently, they were asked which year and month they first used the substance.[13]
- The questions on frequency of use (number of days used in the past twelve months) were expanded from a single question with ranges of days as response options, to a series of alternative questions that determine the actual number of days used.
- Questions on nonmedical use of prescription drugs were revised, including updated, reorganized lists of drugs.
- Several types of data quality checks were embedded in the CAI questionnaire. For example, an out of range response, such as using on 370 days in the past twelve months, triggered an error message, requiring the respondent to reenter a valid answer. Some unusual responses, such as trying heroin for the first time at age 3 or having 200 alcoholic drinks on an occasion, caused a follow-up question asking the respondent if they really meant to give that answer. Some responses that were inconsistent with prior answers triggered an inconsistency resolution, in which the respondent was told their answer conflicted with an earlier response. The respondent was asked which answer was correct.

A Thousand Interviewers in Seven Thousand Places: What Could Go Wrong?

The redesign in 1999 was more than merely a shift to a bigger, better NHSDA. It was the start of an entirely new survey. We should have changed the name of the survey in 1999 to reflect this. In 1998, the survey employed about 300 interviewers to obtain 25,000 interviews. The initial design for 1999 required about 1,200 interviewers to obtain 90,000 interviews (20,000 with PAPI and 70,000 with CAI). Approximately 87 percent of the interviewers working on the CAI version had never worked on the NHSDA prior to 1999. 40 percent had no interviewing experience on any survey. To manage this huge, widespread data collection effort, the number of field supervisors[14] was increased from fifteen in 1998 to eighty in 1999, and the majority of these supervisors had no NHSDA supervisory experience. Experienced, savvy field staff are a key factor in achieving a high response rate. They have critical skills that aid in making initial contacts, explaining the importance of the survey, and convincing reluctant respondents to agree to be interviewed.

Early in 1999 RTI recognized the preparations and planning for the 1999 data collection had been inadequate. In an April management meeting, RTI reported to SAMHSA that of the 1,282 interviewers trained in the first week of January, 232 had left the project. Fifteen field supervisors had also left. Data collection was lagging. It looked like response rates were going to be low if immediate actions were not taken. It was not just the sheer size of the sample that made it difficult to complete all of the data collection on time. Essentially, there were two separate surveys – the national sample of 20,000 for the old (PAPI) questionnaire and the state-based sample of 70,000 interviews for the new (CAI) instrument. The sample of 7,200 segments for the CAI sample covered every region of every state, all year long. In many remote areas, it was not easy to find qualified, committed interviewers to complete the assignments. Weather conditions hampered efforts in some areas.

SAMHSA and RTI agreed on a plan to address the data collection problems. Adding several million dollars to the cost of the project, the plan included:

- Decreasing the workload for interviewers who preferred to work only a few hours per week.
- Holding extra training sessions tailored to the needs of specific regions.
- Increasing the management-to-staff ratio.
- Increase pay rates for interviewers.
- Setting up a team of "traveling interviewers" to send to problematic areas as needed.
- Splitting each quarterly sample into partitions and assigning to FIs as needed to control work flow.
- Conducting exit interviews to find out why interviewers are leaving the project.
- Carrying out analysis to understand reasons for the decline in response rates.
- Reducing the PAPI sample by 5,000 interviews.

Although the response rate improved throughout 1999, the end result was a significantly lower response rate in 1999 compared with prior years. The weighted screening response rate was 92.9 percent in 1998, and dropped to 89.6 percent in the 1999 CAI sample.[15] The interview response rate dropped from 76.0 to 68.6 percent. In the midst of the data collection difficulties during 1999, fixing the PAPI sample was a lower priority than the CAI sample. Thus, the screening and interview rates for the PAPI sample were even lower (83.8 and 66.6 percent, respectively). The final sample sizes were below the planned numbers: 66,706 for CAI and 13,809 for PAPI.

Low response rates raise the possibility that nonresponse bias affects the survey's estimates. For the redesigned NHSDA, the concern was not only with national estimates but also state estimates. Some states had very low response rates, while others had high rates. There was concern that nonresponse bias could impact state-by state comparisons. The interview response rate was below 60 percent in Delaware, Connecticut, and New York. But it was over 80 percent in other states (Mississippi, Arkansas, Utah).[16] If there is a high rate of drug use among nonrespondents, states with lower response rates would appear to have lower rates of drug use than was true. For example, the difference in response rates might cause Utah and New York to appear to have similar levels of drug use, even though the true rate is higher in New York. Nonresponse adjustments built into survey weights may correct for this bias. But without accurate information on the drug use rates of the nonrespondents, one can never be certain how well the nonresponse adjustments reduce the bias. One possible solution to potentially "equalize" the nonresponse bias across states was to target extra resources (more staff, more follow-up visits to households to obtain cooperation) towards states with low response rates. That approach was troublesome because it in effect was applying different methods across states. These were some of the considerations we faced in managing data collection, constructing analytic weights, and interpreting results for the 1999 and later NHSDAs.

Respondents Report More Drug Use with Inexperienced Interviewers

Another impact of the inexperienced field staff was discovered. The analysis of the two simultaneous nationally representative samples (CAI and PAPI) revealed that the level of experience of an interviewer was correlated with respondents' reporting of substance use. Interviewers with more experience tended to obtain lower rates of substance use among the respondents they interviewed. Evidence of this phenomenon had been seen in the 1990 NHSDA field test,[17] and it was confirmed with the much larger sample in the 1999 survey. Given this "interviewer effect," the plan to use the dual sample analysis to create a bridge between the new and old survey designs had to be reconsidered. With the influx of new interviewers, the 1999 data (both PAPI and CAI) were apparently not comparable to the 1998 and earlier data. We began a new research project to attempt to find a statistical model that could provide a way to assess long-term trends, comparing pre-1999 estimates (PAPI) with estimates from 1999 and subsequent years (CAI). A preliminary model was developed that produced adjusted 1999 PAPI estimates for a small set of key indicators, for comparison with 1998 data. But a satisfactory adjustment to broadly permit valid trend assessment was not found. The simultaneous impacts of interviewer effects, mode change (PAPI to CAI), and the dual samples were too complex. We were able to determine that the interviewer effects were smaller with ACASI than with PAPI, but the underlying cause for higher substance use rates with new interviewers was not identified. There was evidence that one factor was shortcuts in protocols that experienced interviewers had learned. To counter this possibility, interviewer training and monitoring was later enhanced to emphasize strict adherence to the specified procedures to promote consistency across interviewers and interviews.[18]

The Mode Effect on Reporting of Drug Use: Paper versus Computer

The interviewer effect limited the usefulness of the 1999 dual sample data for preserving trend comparisons. However, we were able to study how the mode change affected drug use estimates.[19]

A pure mode effect was difficult to measure because of the inherent structural differences between the modes. The PAPI answer sheets had no skip patterns and the ACASI did. This impacted editing and imputation. Comparisons of the unedited lifetime use variables were of interest because they best showed a clean estimate of the mode effect. Skip patterns do not affect this comparison because the ACASI skips occur after the lifetime question for each substance. Unedited lifetime use estimates were consistently higher with ACASI than PAPI. Comparisons of the final edited, imputed estimates of past month drug use showed mixed results, with ACASI giving higher rates for some drugs and PAPI giving higher rates for some drugs. Given CDC's insistence on using ACASI for the tobacco module, it's interesting to note the rate of past month cigarette use among youth was higher with PAPI than ACASI data. One clear conclusion from the mode study is that the CAI mode of interviewing leads to more internally consistent and complete data than the PAPI mode. Thus, the switch from PAPI to CAI reduced the need for and impact of editing and imputation.

Release of the 1999 State and National Results

In planning for the release of the 1999 survey results, a consideration was the volume of estimates to be released. One possible approach was to first release the national data as in prior years, and release the state results later in a separate report, since it took more time to produce the state estimates because of the modeling it entailed. That idea was nixed. HHS officials were adamant that national and state estimates must be combined in one report and released in late summer. The redesigned NHSDA was a legacy accomplishment of Donna Shalala and the administration. It was essential that the first state estimates be released before they left office in January 2001.

We chose twenty specific measures to produce via small area estimation methods for each state. But the work for all twenty measures could not be completed in time for the late August release date. So we chose a subset of seven measures of most interest that would be included in the initial report, Summary of Findings.[20] The other thirteen would be released in December. We also decided to present state comparisons using colored maps showing broad groupings. For each measure, states were ranked from highest to lowest rate. Five approximately equal sized groups (quintiles) were formed. The first group, colored in red on the maps, consisted of the approximately ten states with the highest rates. The report referred to the maps and described the states in the "highest category" and "lowest category." This presentation helped avoid detailed comparisons of individual states with similar rates, where no differences would be statistically significant. While the approach did help avoid over-interpretation, all of the state estimates were available in tables, and some discussion of specific state differences was unavoidable. There has always been interest in identifying the two states with the highest and lowest rates, even though the estimates for these states were never significantly different from the second highest and second lowest states, respectively.

Prior to the release, a briefing for Secretary Shalala and other HHS leaders was held at HHS headquarters. There was obvious enthusiasm and anticipation in the room as we discussed these initial state-level results. Shalala, repeating the advice of our consultant Wes Schiable, warned against over-interpretation based on the first year of the redesigned survey. Several years of data would be needed before patterns and trends would become clear. Excited about the potential of this new tool for tracking drug use, Shalala ordered the survey be renamed. After several ideas were suggested by attendees at the meeting, we said we would work on it and change the name as soon as it was operationally feasible. Some of the findings were as expected, such as the high rates of cigarette use in tobacco-producing states Kentucky and North Carolina.

But unexpectedly, Delaware and Massachusetts stood out with high rates of illicit drug use. Shalala asked us to contact officials in these two states and discuss the results before the press conference. This was partly to give the states advance notice, but also to give us a "check" on our results. A contact in Massachusetts was not surprised by the NHSDA findings. They already believed their state's drug problem was worse than the rest of the country, and appreciated confirmation of this by the federal government. Delaware health officials were skeptical, saying in news articles that "In other reports, Delaware fares much better." After the release, we continued a dialogue with Delaware analysts and conducted further analysis to help understand the high rates in Delaware.

The press conference was held on August 31, 2000. Shalala and McCaffrey emphasized the continuing decline in youth drug use shown by the survey. Shalala said "we have miles to go in our journey to a drug free America." She added "That's why, in 1996, I challenged my department to develop a whole new approach to fighting substance abuse. I wanted our efforts to be based on science, with measurable outcomes, and designed to help community and state leaders formulate targeted programs. As a result, we have greatly expanded the 1999 National Household Survey." Shalala described how the expanded survey can provide "parents, governors, and future administrations" with data to form policies and track progress. She sent letters with a copy of the "Summary of Findings" report to all fifty governors and the mayor of the District of Columbia to apprise them of the NHSDA estimates for their jurisdictions. At least one of the governors sent a thank you reply back to Shalala. Texas governor and Republican presidential nominee George W. Bush wrote back a month later, saying "I share your concern about illegal drugs, alcohol and tobacco, and believe that prevention must be a priority for state and federal governments. I have given the material to my policy staff for review."[21] Four months later, Bush was sworn in as the 43rd president of the United States.

Notes

1 SAMHSA, *Development of Computer-Assisted*.

2 Outside consultants: Jim Anthony Johns Hopkins University), Mick Couper (University of Michigan), Pat Ebener (RAND Corporation), Denise Kandel (Columbia University), Bill Kalsbeek (University of North Carolina), Phil Leaf (Johns Hopkins University), Nancy Mathiowetz (University of Maryland), Peter Reuter (University of Maryland). Federal government staff outside OAS: John Carnevale (ONDCP), Marilyn Henderson (CMHS), Lana Harrison (NIDA), Art Hughes (NIDA), Ron Wilson (NCHS). OAS staff: Don Goldstone, Anna Marsh, Deborah Trunzo, Al Woodward, Joe Gfroerer, Peggy Barker, Joan Epstein, Janet Greenblatt, Joe Gustin, Doug Wright.

3 These were regularly held regional meetings sponsored and chaired by OAS, to discuss collaboration on data issues (primarily treatment data) with groups of states.

4 Memo from Bill Corr, Chief of Staff, to David Garrison, Principal Deputy Assistant Secretary for Planning and Evaluation. State Substance Abuse Data. February 1997.

5 HHS officials included Secretary Shalala, Deputy Secretary Kevin Thurm, Chief of Staff Bill Corr, SAMSHA Administrator Nelba Chavez, CDC Director David Satcher, NIDA Director Alan Leshner, and several others.

6 House of Representatives, Conference Report to accompany H.R. 2264.

7 Chavez, Report to Congress, Conference Appropriations Committee.

8 This statistic was often cited in statements about the magnitude of the youth tobacco problem. It was derived from NHSDA incidence estimates. For example, with an estimated 1.1 million youths under age 18 using cigarettes for the first time per year, dividing by 365 results in about 3,000 per day, on average.

9 Federal Register, Vol. 63, No. 120, 34191–92.

10 The original panel included statisticians Robert Groves (University of Michigan), Michael Hidiroglou (Statistics Canada), and William Kalsbeek (University of North Carolina); drug abuse survey expert Patrick O'Malley (University of Michigan); tobacco epidemiologist Gary Giovino (Roswell Park Cancer Institute); alcohol researcher Raul Caetano (University of Texas); prevention researchers Michael Arthur (University of Washington) and Barbara Delaney (Partnership for a Drug Free America); mental health researcher Philip Leaf (Johns Hopkins University). Other consultants who became members of the expert panel were Statisticians Graham Kalton (Westat) and Alan Zaslavsky (Harvard University); drug policy experts Peter Reuter (University of Maryland) and John Carnevale (Carnevale Associates).

11 The panel met for the last time on November 6, 2008, shortly after Peter Delany was appointed OAS Director.

12 A supplemental sample of 2,500 youths was added in years 1999 and 2000, with funding provided by HHS to support the tobacco module.

13 If the reported age at first use was the same as or one year less than the respondent's current age, then the month of first use question was asked. This approach guaranteed that month of first use would be obtained for all past year initiates.

14 Each field supervisor was responsible for managing a team of about ten to fifteen interviewers in a specific territory, such as a state.

15 Beginning with the 1999 survey, SAMHSA computed weighted response rates for use in published descriptions of the survey results.

16 SAMHSA, *Summary of Findings 1999*, C-13.

17 Turner, Lessler, and Gfroerer, *Survey Measurement*, 218–9.

18 Gfroerer, Eyerman, and Chromy, *Redesigning an Ongoing*, 161–84.

19 Gfroerer, Eyerman, and Chromy, 135–59.

20 The seven measures were past month (1) any illicit use, (2) marijuana use, (3) other illicit drug use, (4) cigarette use and (5) binge alcohol use and past year (6) illicit drug dependence, and (7) illicit drug or alcohol dependence.

21 Letter from George W. Bush to Donna Shalala, September 25, 2000.

10 Continuing Survey Design Improvements

The field staffing difficulties encountered in the 1999 survey expansion were alleviated somewhat in 2000 because the survey no longer included the additional paper-and-pencil interview (PAPI) sample. The 2000 NHSDA was designed to obtain 70,000 interviews, all using computer-assisted interviewing (CAI). Nevertheless, response rates and management of the large, widely dispersed data collection remained a concern. SAMHSA and RTI continued to discuss and explore new approaches, including respondent incentives (cash payments to respondents for completing the interview) and renaming the survey.

Incentive Payments

Serious consideration of using incentives began early in 1999, when the data collection difficulties emerged. SAMHSA and RTI identified the pros and cons, and then discussed it at the first NHSDA Expert Consultant Panel meeting in October. The panel's comments on incentive payments helped shape the next steps. Some of the questions addressed at the meeting were: Will incentives increase response rates? If monetary incentives are used, how much should be paid? Should the incentive be prepaid or promised? Should there be remuneration for all sample members, or only for refusal conversion or only in low-responding states or only in particular sub-populations? Should differing levels of incentives be used? What effect would incentives have on screening participation, quality of responses, and sample distributions? Will drug use reporting be influenced by incentives? Could there be a perception that the government is paying the respondent to admit illegal behavior? How might the media report this? The panel said there was a need for systematic experimentation. They recommended that before SAMHSA begins paying respondents, other options for improving response rates should be explored. The incentive payment approach should be fully tested to show that it would be cost-effective and would improve response rates.

We developed a plan for an incentive field test. It was scheduled for the first half of 2001, so results would be available in time to decide whether or not to implement incentives in the 2002 NHSDA. A random subsample of 9,600 NHSDA respondents participated in the experiment. This included 4,233 who received no incentive, 2,489 who received twenty dollars cash, and 2,878 who received forty dollars cash after the completion of the interview. Measures of respondent cooperation, data quality, survey costs, and substance use rates were compared across the three groups.[1]

Early results from the experiment were reported soon after data collection ended in June. They were encouraging. Response rates were 69 percent in the no-incentive group, 79 percent in the twenty dollar group, and 83 percent in the forty dollar group. The net cost for data collection (i.e., including the incentive payment) was lower in the twenty dollar and forty dollar groups than with the no-incentive group, because sampled persons offered an incentive were more willing to agree to participate when first approached by the interviewer. There was less need for interviewers to make return visits. The reduction in travel costs more than offset the cost of providing the cash. Also, initial analysis showed the incentives did not affect reporting of substance use. These findings led to two decisions. First, since the incentive study subsample was relatively small and the data were consistent with the remainder of the NHSDA sample, all of the field test respondents were included in the data set for producing 2001 NSDUH estimates. Second, OAS Director Don Goldstone, a strong proponent of the use of incentives in the survey, decided we should begin offering incentive payments in the 2002 survey. He chose thirty dollars as the amount to be given to respondents. The reasoning was that the majority of the gain in response rate was found with the twenty dollar incentive. The additional gains with forty dollars were smaller and less cost-effective.

Improvements in Data Collection Quality Control

After discovering that respondents were reporting less substance use when interviewed by experienced interviewers (the interviewer experience effect), we began a program of interviewer observation, to gain a better understanding of how interviewers interacted with respondents. RTI and OAS staff familiar with the NHSDA data collection procedures accompanied a targeted sample of interviewers as they visited households and conducted interviews. The observers took notes on the interviewers' performance, checking a pre-coded list of specific behaviors such as whether or not they read an introductory statement verbatim.

Widespread departures from the protocol were found, especially among experienced interviewers. As a result, in July 2001 new guidelines were sent to all interviewers, and during October and November extra training was given to reinforce the importance of following the procedures exactly as they were specified. It was critical for all interviewers to administer the survey in the same way, to eliminate the interviewer effect that had adversely affected the 1999 results. New lessons were added to the face-to-face veteran interviewer training in the first week of January, for the 2002 survey. Finally, a certification process began in the January 2002 training, in which interviewers would not be allowed to conduct interviews until they demonstrated they were able to conduct interviews correctly. Later studies did indicate that the interviewer effect gradually diminished in subsequent surveys as the emphasis on following protocols continued.[2]

Changing the Name of the Survey

In the final months of the Clinton administration, HHS Secretary Shalala told SAMHSA to change the name of the survey. She felt the name should reflect the survey's new design and capability to track substance use in states. But there were other factors to consider when deciding on a new name for the survey. Including both "state" and "national" in the title could result in an unwieldy, confusing name. We wanted the name to be simple, yet accurately portray the survey purpose and content. But the name could be a factor in achieving respondent cooperation. Interviewers had reported that some respondents reacted negatively when they were approached to participate in a survey about their "drug abuse." "Drug use" would be seen as less judgmental than "drug abuse." And since the survey covers tobacco and alcohol, perhaps "substance" might be more appropriate than "drug." Since the survey covered group quarters, "household" was no longer an accurate depiction of the population we were surveying. "Health" in the survey name could help in gaining respondent cooperation due to its salience for everyone, and we expected to be adding more questions about health. With all of these considerations in mind, we asked RTI to carry out an "Evaluation of Potential Name Change" in March 2001. Supervisory, office, and interviewing staff responded to questionnaires and participated in conference calls to give their opinions on new names and terms under consideration. The results were compiled in a report that SAMHSA used to make a final decision. Ultimately, the decision was made by OAS Director Don Goldstone. He liked "National Survey on Substance Use and Health." Almost everyone agreed this was fine. Except ONDCP. They insisted that "drug" must be in the survey name. So the new name would be the "National Survey on Drug Use and Health," starting with the 2002 survey, the same year that we anticipated the incentive payments would begin.

Late in 2001, there was an objection. The new SAMHSA Administrator Charles Curie[3] was upset that OAS was changing the name of the survey. In his view, the survey had name recognition, so it was better to keep the old name. Dr. Goldstone explained the advantages of the new name to Curie, and added that it was too late to go back. RTI had already converted most of the survey materials, and there was not enough time to revert everything back to the old name. Besides, the HHS Secretary had ordered us to change the name. Curie conceded. Actually, we probably could have reversed the decision when Curie complained, but it would have been costly and risky to change survey materials back to the old name with so little time before 2002 data collection would begin. It also may have damaged morale among the data collection staff. And Don preferred the new name. He and most of the OAS and RTI staff working on the NHSDA (including myself) agreed it was better in the long run for the survey.

Concern about Trend Measurement

Although the initial analysis of the field test did not indicate that rates of substance use would be affected by the incentive payments, there were concerns that some small changes in rates of use would become evident when incentives were fully implemented in the large 2002 sample.[4] The name change and improved quality control could also affect response rates and reporting of substance use. But after the 1999 debacle, we were hesitant to add a supplemental sample to "bridge" the old and new data for trend purposes. A dual-sample design with different management practices (none versus enhanced), a different survey name (NHSDA versus NSDUH), and different incentive payments (zero dollars versus thirty dollars) in the two samples did not seem feasible. The cost would be substantial, and it would be difficult to successfully control the experimental factors. Furthermore, we believed any differences in substance use rates caused by these factors would be small and probably not statistically significant. I discussed trend measurement with Goldstone, pointing out the changes to the survey that had the potential of affecting respondents' reporting of substance use in 2002. He was clearly aware of this, but he was committed to the new survey name and the incentive payment. I pointed out that the sample size for the 2001 incentive experiment was small, so it was possible that with incentives used in the full sample of 67,000 respondents, a small increase in drug use rates due to the incentive might be statistically significant, and we would not be able to measure the effect separately from any real change in rates, nor adjust for it to preserve trends. I warned him that ONDCP and others would be upset if there were methodological impacts that made the 2002 data not comparable with earlier years. Don was not concerned. In fact, he seemed defiant. His reply, as I recall, was approximately, "Good. Let those bastards deal with it." I assumed he was referring to either the new SAMHSA leadership, ONDCP, or the Bush administration. But I didn't ask.

Questionnaire Changes

In designing the 1999 questionnaire, one goal was to limit the length. This cautious approach was taken because although the instrument had been thoroughly tested, no survey had ever utilized ACASI on such a large scale, both in terms of the number of respondents (70,000) and the length of the ACASI portion of the interview (roughly a half hour). In addition, the abrupt decision to fully implement the CAI in 1999 instead of gradually over 2000 and 2001, cut short the time for development of supplemental modules. With minimized content and the use of skips within the ACASI questionnaire, the average interview time for the entire CAI in 1999 turned out to be only fifty-two minutes. This was about seventeen minutes shorter than the average interview time for the 1998 paper-and-pencil questionnaire. There was no evidence that respondents were fatigued or unwilling to spend twenty minutes or more completing the ACASI. Thus, there was room to add more questions in 2000 and 2001 (in the non-core part of the interview). By 2001, the average interview time was up to sixty minutes after the addition of modules covering mental illness and other topics.

One useful feature of computer-assisted interviewing is the capability to calculate precise timing of each section of the questionnaire, by inserting time stamps anywhere in the instrument. In fact, detailed analysis of keystroke patterns, such as how long respondents take to answer a particular question or how often they change an answer, can be useful in designing and improving the questionnaire.[5]

Dr. Nelba Chavez, SAMHSA Administrator from 1994 to 2000, had urged OAS to add more mental health questions to the NHSDA. We consulted with CMHS and other mental health experts to determine what specific topics we should cover.[6] Based on their recommendations, new modules asking about youth and adult mental health service utilization were added in 2000. Also, a youth mental health module that consisted of short scales intended to provide a basis for estimating the prevalence of specific mental disorders was added. Additionally, questions needed to capture substance use disorders (abuse and dependence) based on DSM-IV criteria were added. This was necessary for the estimation of drug abuse treatment need and the "treatment gap" according to the definition determined by an ONDCP-chaired interagency workgroup in 1998. Other items added to the survey in 2000 included questions regarding the locations where treatment was received, whether any treatment was being received currently, and the method of payment for current treatment; questions about cigarette purchase prices and about purchases by underage individuals; and questions to ascertain industry and occupation for employed persons.

More changes were made for the 2001 survey. Most had been suggested by the NHSDA Expert Consultant Panel:

- Included a new module with questions about specialty cigarettes such as bidis and clove cigarettes.
- Replaced DSM-based questions about cigarette dependency with a new series of questions on cigarette dependency from the Nicotine Dependence Syndrome Scale (NDSS).
- Added several questions designed to measure respondents' awareness of marijuana laws in their states.
- Included a new module with questions about market information for marijuana, such as cost and amount purchased.

The mental health content of the survey continued to evolve. The adolescent mental health module added in 2000 was dropped in 2001, because there was no acceptable algorithm to convert the data from the short scales into representative population estimates of the prevalence of mental illness among youths. But there was optimism that an algorithm could be developed in which responses to a set of short scales in the questionnaire were combined to produce estimates of serious mental illness (SMI) among adults. These scales were added to the 2001 questionnaire, and the estimating algorithm was determined from a clinical interview study conducted during 2000.[7]

The 2001 NHSDA included a test of the item count method, which indirectly estimated prevalence of sensitive behaviors. Respondents were shown a list of behaviors and asked to answer how many of these behaviors they had engaged in during the past year. For a random half of the sample, the list consisted of four behaviors. For the other half-sample, the list included the same four items, plus one more item – use of cocaine. The average number of items reported in the four-item half-sample was then subtracted from the average number from the five-item half-sample, resulting in an estimate of the prevalence of cocaine use in the past year. The results of this test were not encouraging. The indirect estimate was actually lower than the estimate derived directly from the standard core question on last use of cocaine.[8]

There were no major changes to the questionnaire for the 2002 survey, except for a few minor adjustments to improve clarity and the addition of follow-up probe questions.

The information given to respondents prior to completing the interview changed each year. The opening statement of the household screener in 1999 told the respondent the survey was sponsored by the Substance Abuse and Mental Health Services Administration. That was changed to the Department of Health and Human Services in 2000, and again to the US Public Health Service in 2001.[9]

Collecting Urine and Hair from NHSDA Respondents

Despite the data collection approaches employed in NHSDA to obtain a representative sample and maximize honest and accurate reporting, analysts have assumed that the survey underestimates true drug use prevalence rates. The underestimation is thought to be due to some combination of (1) the sample not capturing drug users (coverage or nonresponse error) and (2) respondents not reporting their drug use (response error). The amount and sources of the underestimation are not known, but the bias resulting from these errors undoubtedly varies by drug and demographic group. After the public criticisms of the survey's heroin and cocaine estimates in the early 1990s, the Government Accounting Office (GAO) recommended that a NHSDA validation study should be done to assess response error.[10] An opportunity to conduct such a study arose in the late 1990s, when NIDA grantee Dr. Lana Harrison contacted me to propose a collaboration. Lana had worked on the NHSDA team in my branch at NIDA, remained at NIDA after the 1992 ADAMHA reorganization, and then took a position as an Associate Professor at the University of Delaware. Familiar with the continuing skepticism of the quality of self-reported drug use data and the GAO recommendation, she had applied for and was awarded a grant from NIDA to conduct a validation study. In this study, survey respondents, after completing the interview, would be asked to provide hair and urine samples that would be tested for drugs. Lana wanted SAMHSA to co-fund her study to increase the sample size to get more useful results. Recognizing the relevance of the study for NHSDA, we agreed to collaborate. Using funds for methodological work included within the NHSDA contract, we arranged to embed her study, with an increased sample size, in the 2000 and 2001 NHSDAs.

NHSDA interviewers were canvassed and asked if they would be willing to collect the specimens as a part of their regular interviewing assignments. Most were willing. During the January 2000 veteran[11] interviewer training, the volunteering interviewers received special training on how to introduce the follow-up study, how to cut a snippet of hair from the respondent's head, how to collect the urine sample, and how to package and mail the specimens to the testing laboratory. The study was not mentioned to respondents until they had completed the interview, so that their answers would not be influenced by knowledge of the test. A sample of 4,465 respondents age 12–25 years who had completed the NHSDA interview were asked to participate. Each person was offered a cash incentive of twenty-five dollars for each specimen.

The study was successful in getting respondents to agree to provide the specimens. About 85 percent allowed the interviewer to cut a piece of their hair, and about 85 percent agreed to provide a urine sample. Most respondents reported their recent drug use accurately, but the study findings were limited. Due to technical problems associated with the hair testing, the analysis focused primarily on the urine tests. There was generally strong agreement between self-reports and urine test results, but there were some respondents who reported use and tested negative, and others who reported no use and tested positive. The results demonstrated the difficulty in attributing a specific time period of use to a positive drug test. The tests could not show precisely how recently a person had used a drug. Factors that may affect whether a drug user shows a positive urine or hair specimen include the recency of use, quantity used, number of times used, potency of the drug, environmental exposure, hair color and treatments, and the cutoff level chosen by the laboratory. Also, the numbers of cocaine and heroin users in the sample were small, making it difficult to draw conclusions about reporting of these drugs.[12]

Results from 2000 through 2002

The release of the 2000 NHSDA results was abbreviated, low-key, and late. This was mainly due to the September 11, 2001 terrorist attacks. A brief press release on October 4, 2001 announced the NHSDA findings. The focus was on a decline in drug use among 12- and 13-year-olds and the importance of parental involvement in keeping kids off drugs. The SAMHSA report released on that day included national results only.[13] The state-level results would be in a separate report to be published at a later date. Also missing from the initial report was the data on substance use disorders, treatment need, and treatment. President George W. Bush, in one of his first drug abuse policy initiatives, had announced on May 10 that he had directed HHS Secretary Tommy Thompson to "conduct a state-by-state inventory of drug treatment need and capacity, and to report back to him within 120 days on how to most effectively close the treatment gap in this country." It was decided that all of the new NHSDA data related to this issue would be kept out of the annual national findings report, and instead would be presented in the treatment gap report to the president. The report would rely heavily on the new national and state treatment need estimates based on the substance use disorder module added to the 2000 questionnaire.[14] Bush's 2003 National Drug Control Strategy, released in February 2003, cited the NHSDA data and called for increases in funding for treatment, through a new initiative called Access to Recovery:

> The vast majority of the millions of people who need drug treatment are in denial about their addiction. Getting people into treatment – including programs that call upon the power of faith – will require us to create a new climate of "compassionate coercion," which begins with family, friends, employers, and the community. Compassionate coercion also uses the criminal justice system to get people into treatment. Americans must begin to confront drug use – and therefore drug users – honestly and directly. We must encourage those in need to enter and remain in drug treatment. The President's National Drug Control Strategy envisions making drug treatment available to many more Americans who need it.
>
> Overall, for 2003, the Administration proposes $3.8 billion for drug treatment, an increase of more than six percent over 2002. This includes a $100 million increase in treatment spending for 2003 as part of a plan to add $1.6 billion over five years. Getting treatment resources where they are needed requires us to target that spending. This budget asks that $50 million of new treatment funding be targeted to areas with greatest need.[15]

The release of the 2001 NHSDA national results began a new approach that would continue for at least sixteen years. September marked the celebration of National Alcohol and Drug Addiction Recovery Month. To kick off Recovery Month, a press conference was held during the first week, and SAMHSA decided to tie that event to the release of the survey results. Not only would this help to stabilize the data release timing each year, but it would help garner attention to Recovery Month. The press conference was held on September 5, 2002. Speakers included ONDCP Director John Walters, HHS Secretary Tommy Thompson, and SAMHSA Administrator Charles Curie. The survey showed increases in past month use of marijuana, cocaine, and nonmedical use of prescription drugs between 2000 and 2001.

During the preparation of the report, unusual and troubling results were found. There were surprisingly large increases in lifetime use of some drugs, including marijuana. Lifetime use estimates for the full population should change according to predictable patterns from year to year, since the only additions are new initiates and immigrant users, and the only decreases are from deaths among lifetime users. The size of these groups can be reliably estimated each year based on prior years' data. For marijuana, previous studies showed fewer than three million new users each year, and relatively small numbers of immigrant users and deaths among users. But the 2001 lifetime marijuana estimate was seven million more than the corresponding 2000 estimate. An investigation was conducted by RTI to determine if the incentive experiment, embedded within the main 2001 data collection, or the enhanced data collection quality control could have affected the estimates. The conclusion was that there was some effect, but it was small. Increases in past month use were deemed valid. A discussion of the investigation was included in the report.

The concern shifted to the 2002 data, and the investigation continued. But the analysis was more complicated due to the survey name change and full implementation of incentives. After the first half of the 2002 data was collected, processed, and tabulated, it was clear the survey changes had affected the data. OAS Director Goldstone informed ONDCP there was a possibility that valid comparisons between 2002 estimates and estimates from prior years may not be possible. ONDCP Director Walters was livid. His boss, President Bush, had told him there were two main goals the National Drug Control Strategy must have: Reduce the rate of current drug use among adults, and reduce the rate of current drug use among youth. These two goals were stated in Bush's first drug control strategy, published in February 2002. The NSDUH was to be the source for tracking these two goals, and the baseline year was 2000, the year before Bush took office. Now, just as ONDCP was preparing the 2003 strategy, SAMHSA was telling them the data could not be used this way. ONDCP demanded an explanation and suggested an outside expert review. And they wanted a memo from SAMHSA documenting the evidence and SAMHSA's recommendation.

We brought a group of survey methodology experts together to advise us on the data concern.[16] A representative from ONDCP attended. The expert panel concluded that the changes in the survey methods did affect the comparability of 2002 data with data from prior years, and the changes improved the quality of the data. They recommended against applying any kind of adjustment to the estimates to permit comparisons across survey years. There was not enough information to create a valid adjustment.

Don Goldstone sent a memo to ONDCP explaining the data problem and inability to adjust the data, along with a summary of the expert consultant meeting. The 2002 estimates would constitute a new baseline. There should be no comparison of 2002 data with prior years. The memo closed with "SAMHSA does not at this time anticipate further changes of the NSDUH in the near future."[17]

Accepting this conclusion, ONDCP decided to resort to 2002 as the baseline year for tracking the rate of adult drug use. There was no alternative to NSDUH. But for tracking youth drug use, there was another data source that could provide an earlier baseline. Monitoring the Future (MTF) surveyed eighth, tenth, and twelfth graders each year. MTF Principal Investigator Dr. Lloyd Johnston was asked if he could construct a new type of estimate, combining the samples from the three grade levels to create a single estimate to portray youth drug use. It was not ideal, but Walters was adamant about having the earlier baseline. Johnston agreed it was feasible and promised to provide these estimates. ONDCP decided to use 2001 as the base year for the new MTF estimates. The strategy with these two goals with different data sources and baseline years was released in February 2003.

SAMHSA released the 2002 NSDUH data at the September 5, 2003 Recovery Month kickoff press conference. The report and the press conference focused just on the 2002 data.[18] The only comparisons with earlier years were with the retrospective incidence and lifetime use estimates, which were all computed from reports of age at first use by respondents in the 2002 survey. The report included extensive discussion of the methodological changes and the investigation into their effects. Also, there were prominent warnings to readers, such as:

> Because of improvements to the survey in 2002, estimates from the 2002 NSDUH should not be compared with estimates from the 2001 and earlier NHSDAs to assess change over time in substance use. Therefore, the 2002 data will constitute a new baseline for tracking trends in substance use and other measures.

At the press conference, Curie put a positive spin on the situation, saying it was a transitional year for the survey, and "this improved survey sets a new standard for describing substance use." The headline of the press release was "22 Million in U.S. Suffer from Substance Dependence or Abuse" and the president's Access to Recovery initiative was highlighted. Walters pointed out the "denial gap," referring to the NSDUH statistic that among those who needed but did not receive treatment for a drug problem, as determined by the symptoms they report on the survey questionnaire, 94 percent did not feel they needed treatment. HHS Secretary Thompson did not participate in the press conference.

Impact of the Changes in 2002

Although it's not possible to quantify the effects of each of the survey changes, the consensus among the SAMHSA and RTI NSDUH teams was the incentive payment was the major factor. The effects were (1) an increase in the response rate, (2) an increase in reporting of substance use, and (3) reduced data collection costs. Overall, the interview response rate (weighted) increased from 73.3 percent in 2001 to 78.6 percent in 2002. But the improvement differed by age. For youth, the rate improved by eight percentage points (from 82.2 percent to 90.0 percent). The improvement was even greater among young adults age 18–25 (75.5 to 85.2). Older adults have consistently had the lowest response rate, and the incentive did little to improve it (69.9 to 71.5 percent).

An analysis of patterns of changes in drug use rates from 2001 to 2002 showed that increases in rates of lifetime use of several drugs were unrealistic.[19] For example, the estimated number of lifetime marijuana users increased by more than ten million, whereas estimates of new users per year had been less than three million. The analysis also assessed the potential effect of the increased response rates, considering the possibility there was a high percentage of drug users among the additional 5 percent of the sample swayed by the thirty dollars to participate. The analysis showed that the additional respondents could not account for the increase in estimated use. The conclusion was that the incentive probably affected how people answered the questions on drug use. In contrast with the impact on response rates, the increase in substance use reporting was substantial for adults, but small for youth.

The incentive made the interviewers' jobs easier. They were more confident and found that reluctant respondents were more willing to agree to participate if they knew they would get thirty dollars. Fewer trips were needed to revisit sample addresses. This translated to a cost savings that exceeded the $2 million given out to respondents as incentives. Additionally, assignments were completed much more quickly. In fact, we later became concerned that the data collection was no longer spread evenly throughout each quarter. A large portion of the interviewing was done in the first months of each quarter. The increase in response rates and the reduced travel led RTI to conclude that significant reductions in field staff could be made without jeopardizing response rates. However, with the incentive payments there was a greater concern that rogue interviewers might falsify interviews (curbstoning), not only charging hours for work not done, but pocketing the thirty dollars. RTI developed enhanced systems for rooting out falsification. Nevertheless, it occasionally does occur.[20]

Four Years of Redesign

Between 1999 and 2002, the survey was transformed. The 1999 redesign was a planned improvement to the survey, and was expected to affect estimates and comparability with prior years. But in hindsight, it actually took another three years to finish the job. Improvements in managing the data collection were implemented throughout the period, especially beginning in the second half of 2001. Questionnaire enhancements were made every year. Introductory information given to respondents by the interviewers changed every year. After the name change and incentive introduction in 2002, it seemed the survey overhaul had finally reached its completion. It had to. People were frustrated with the lack of consistency in the survey methods and the inability to ascertain trends from the data. No more changes to the design causing breaks in trend would be tolerated. There would be at least thirteen years of consistency beginning with the 2002 survey. The 2002 estimates became the new baseline.

Notes

1 Kennet and Gfroerer, *Evaluating and Improving Methods*, 7–17.
2 Wang, Kott, and Moore, *Assessing the Relationship.*
3 Charles Curie became SAMHSA Administrator in November 2001.
4 A more in-depth analysis of the field test completed several months later did find that the incentive was significantly associated with increased reporting of marijuana use.
5 Kennet and Gfroerer, 105–48.
6 A consultant meeting was held November 7, 1997. Consultants: Adrian Angold, James Anthony, Karen Borden, Bill Narrows, Kathryn Rost, Kimberly Hoagwood, Charles Kaelber, Denise Kandel, Ronald Kessler, Philip J. Leaf, Kathleen Merikangas. OAS staff: Peggy Barker, Joan Epstein, Joe Gfroerer, Donald Goldstone, Janet Greenblatt, Anna Marsh CMHS staff: Marilyn Henderson, Ronald Manderscheid.
7 Kessler et al., "Screening for Serious Mental Illness."
8 Kennet and Gfroerer, 149–74.
9 SAMHSA and other health agencies in HHS were components of the Public Health Service, an Operating Division within HHS, under the direction of the Assistant Secretary for Health. Decisions on which organization was named in survey materials were often intended to maximize familiarity and positive reactions from potential respondents.
10 GAO, *Drug Use Measurement.*
11 "Veteran" interviewers are those that had worked on the survey prior to December.
12 Harrison, Martin, Enev, and Harrington, *Comparing Drug Testing.*
13 SAMSHA, *Summary of Findings From the 2000 NHSDA.*

14 US DHHS, *Closing the Drug Abuse.* OAS later published a complete report on the 2000 treatment gap data: SAMHSA, *National and State Estimates* .
15 The White House, Office of the Press Secretary. Fact Sheet on the President's National Drug Control Strategy, February 12, 2002.
16 Consultants participating in the September 12, 2002 meeting were Floyd Fowler, Robert Groves, TIm Johnson, Nancy Mathiowetz, Clyde Tucker, and Alan Zaslavsky. Fe Caces from ONDCP attended, along with OAS and RTI staff.
17 Don Goldstone, memorandum to Terry Zobeck, November 4, 2002.
18 SAMSHA, *Results from the 2002.*
19 SAMSHA, 107–37.
20 The most serious falsification case was discovered in 2011. An interviewer in Pennsylvania had falsified hundreds of cases over six years. This interviewer's "completed" cases were all removed from the NSDUH data files, and data were re-weighted. Although the estimates for Pennsylvania were seriously affected, the impact on national estimates was very small. The cost of these lost cases and the effort to correct the data files was enormous, and was mostly covered by RTI.

Wrap-Up for Chapters 9 and 10

The design of the NSDUH and the reporting of the results have been influenced by various societal, political, personal, and scientific factors throughout its history. Many of these factors emerged and interacted simultaneously in the time leading up to the 1999 redesign. The redesign planning was complex enough at the start, when the NSDUH team was just dealing with the conversion from PAPI to CAI. The NSDUH methodological studies and the research done by others made it clear that CAI would result in more efficient data collection and processing, and better quality data. The early implementation plan, including a supplemental sample of PAPI interviews to provide a bridge for trend analysis, seemed straightforward. And there was no particular pressure to have it fully implemented by 1999. But when the administration decided to expand the survey to produce state estimates, and then to satisfy the requirements of the pending tobacco legislation, those dreams of a statistically clean transition were shattered. Those added requirements created complexity and a new sense of urgency. The state design and estimates were needed before the Clinton administration left office, as they wanted to take credit for the survey improvements. That meant that a report on state estimates needed to be ready before the end of 2000, which meant the expanded sample had to be fielded in 1999. In addition, youth smoking rates, estimated separately for different cigarette manufacturers, were needed as soon as possible to comply with the pending tobacco legislation. Would our plans for implementing and measuring the effects of the PAPI to CAI conversion be compromised by these additional survey changes? We optimistically assumed not. After all, the administration was familiar with the survey and had confidence we could successfully accomplish the redesign. But as it turned out, statistical rigor could not overcome the practical realities and the enormity of the redesign. Our carefully designed random split of the sample to allow us to estimate the difference in reporting between CAI and PAPI was negated by the uncontrolled impact of the huge sample increase. The difficulties in recruiting qualified interviewers and getting them to efficiently complete their assignments led to reduced response rates. More surprising, and probably more consequential, was the discovery that new interviewers, inexperienced on the NSDUH, for some reason produced different rates of drug use reports in the respondents they interviewed compared to interviewers experienced on the NSDUH. The large influx of new interviewers therefore affected the estimates, potentially rendering them incomparable to estimates from previous years.

The discovery of these problems and assessment of their potential impact on estimates of substance use (both national and state level) involved collaboration of the full NSDUH team. The impacts cut across all aspects of the project: recruitment and training of field staff, sample design, data collection, and estimation. Similarly, development of options to fix these problems and to determine how to report the results of the 1999 survey had to be a multi-disciplinary activity with strong coordination between OAS and RTI. To counter the need for additional field interviewers and increased response rates, sample adjustments, ramped up recruiting, and revisions to management and training protocols were implemented. Later, incentive payments to respondents were also added, and the name of the survey was changed.

Extraordinary efforts were made to create statistical models that would estimate the effects of all the survey changes on estimates. The results of these efforts helped to explain the impact of the changes, but were not accurate enough to create a bridge between the old and new data. Comparability with prior survey estimates was lost. This did not help to enhance the statistical integrity of the survey or promote confidence in the NSDUH team. Full documentation of all of the modeling efforts, and an expert panel review and concurrence with our conclusions did help to alleviate the loss in credibility. Also, there was no disputing the fact that the survey design and data quality had been greatly improved for the long run.

The reputation of the survey during the decade after the 1999 redesign was strengthened by the creation of the standing expert consultant panel. The involvement of these highly respected statisticians and researchers helped the NSDUH team to make sound decisions about the survey design and other aspects of the survey. Also, the existence of this panel often swayed others to support our decisions and approve plans we proposed regarding the management of the project. We took advantage of that by mentioning the panel's involvement whenever we thought it would be helpful, such as when we explained to ONDCP that we could not justify using statistical models to construct valid trends after a change in the survey design.

11 Analytic Bankruptcy, Reorganization, Recovery, and Resilience

For a while after the release of the 2002 survey results, the Office of Applied Studies (OAS) endured complaints and criticisms over the break in trend caused by the changes to the survey. Eventually the new baseline was widely accepted, although it was frustrating to many and ignored by some analysts who continued to compare estimates from 2002 and later to data from earlier surveys, to describe long-term trends. OAS NSDUH reports included strong warnings against these comparisons and focused on the 2002 baseline.

After 2002, we were vigilant in keeping the survey methodology consistent, to prevent any more breaks in trends. All major substance use indicators were comparable from 2002 to at least 2014. To outside data users it may have seemed as if the survey was running smoothly, with little disruption or intervention. But in OAS, shifts in management, policy changes, and communication problems created turmoil that affected the survey and the people who worked on it.

SAMHSA's Data Strategy and National Outcome Measures

The legislation that created SAMHSA in 1992 required the agency to conduct evaluations of prevention and treatment programs, to improve the effectiveness of these efforts. But after ten years and billions of dollars provided to states through its block grants, the agency had done little to demonstrate that the funds had made an impact. The George W. Bush Administration advocated the use of performance measures to promote accountability for public services. Towards that end, the administration began a process to identify federal programs that were not working, and eliminate them. The Office of Management and Budget (OMB) created the Program Assessment Rating Tool (PART) in 2002, to rate all federal programs on their effectiveness. In 2003, OMB rated the SAPT Block Grant ineffective, a designation given to fewer than 5 percent of federal programs. SAMHSA needed to take action to address this rating.

The agency needed data to demonstrate that treatment and prevention services supported by the block grant funds were helping states' efforts to prevent and treat substance abuse.

SAMHSA Administrator Charles Curie's solution was a National Outcome Measures (NOMS) program. NOMS would consist of a limited set of key outcome indicators that would be tracked and reported to SAMHSA by state offices applying for block grant funds, and other SAMHSA grantees. Curie wanted all of SAMHSA's data collection activities, including NSDUH, to be aligned with NOMS. The goal was to identify a few critical indicators that all grantees and all SAMHSA surveys would produce. The thinking was that focusing on just these indicators would streamline data collection and reporting and enhance comparability. In conjunction with the NOMS implementation, SAMHSA began to develop a data strategy. The strategy would lay out a long term plan for coordinating and guiding decisions on all of SAMHSA's data. Curie relied heavily on Stephanie Colston, his key advisor, to develop the NOMS and the data strategy.

After OAS Director Don Goldstone retired in February 2004, Goldstone's deputy, Dr. Charlene Lewis, became acting OAS director. OAS staff were concerned that SAMHSA would soon make major cuts to OAS data collection programs or even dismantle OAS and redirect its budget to NOMS-related activities. Some leaders in and outside of SAMHSA, including Curie, complained that OAS had become too academic, focusing on research studies that did not directly contribute to SAMHSA's mission, and that OAS did not share its data or provide analyses helpful to SAMHSA's programs in the three centers. There were increasing demands for center access to NSDUH data files.

Although NSDUH and TEDS were recognized as potentially significant sources of data for some of the NOMS, at first OAS played a minor role in the development of the NOMS program. SAMHSA's goal was to get states to "own" the measures and data that would be used, since states were going to be reporting the results to SAMHSA each year. Despite their statistical expertise and knowledge of key data sets, OAS staff were excluded from much of the initial NOMS development activity. On one occasion, Doug Wright, the OAS statistician responsible for NSDUH state estimates, was allowed to attend a SAMHSA NOMS planning meeting with state representatives, but only if he promised not to say anything. Communication was one way. Input from OAS statisticians was suppressed.

In mid-2005, Dr. Javaid Kaiser was appointed director of OAS. He had been working with Colston on NOMS development in another SAMHSA office. Although Kaiser had a PhD in statistics, survey research was not his area of expertise, and he had little management experience. His role as OAS Director was to help implement Colston's plans for NOMS and the data strategy. Under Kaiser, support for NOMS became the OAS priority. Critical staff responsible for the development of NSDUH questionnaire items and field operations were reassigned to work on NOMS. Research studies by OAS analysts using NSDUH and other data sets were no longer supported, nor were methodological studies. Kaiser ordered a complete review of the NSDUH questionnaire, to identify which items were essential for NOMS or other federal reporting requirements, so everything else could be eliminated. Kaiser and Colston attended large meetings with state and HHS leaders, making presentations describing SAMHSA's new data strategy, claiming that NSDUH had "too much precision" and therefore almost half of the questionnaire could be cut.[1]

I sent memos and met with Kaiser to explain the short-sightedness of these policy and management shifts. NSDUH was a major component in the federal statistical system, with many federal, state, and local agencies relying on its data. SAMHSA was considered the steward of this important survey. In addition, a strong analysis capability was essential to the success of OAS and SAMHSA. Why spend $40–50 million a year collecting data that will sit unused in a data file? Cost savings from removing non-essential questions would have been small, because the length of the questionnaire has a relatively small impact on the survey cost. Sample size is the primary driver of cost in household surveys. It determines the number of interviewers to be hired and trained, and the travel expenses and wages for contacting households. Furthermore, major questionnaire changes would impact trend measurement. I also explained that, unlike the NSDUH public-use file, the NSDUH restricted-use files could not easily be given to the centers or others external to the project. Access was restricted by the laws governing the collection and dissemination of these data. These laws are intended to protect the confidentiality of survey respondents. The data needs of the three centers could only be met through collaboration between OAS analysts and center staff. My arguments were misunderstood and ignored.

The outcomes of the OAS policy and management shifts were easily quantified. Morale plummeted and many staff left. Most of the NSDUH analysts found other jobs. My division lost half (three) of its analysts. The entire Division of Analysis staff was gone by mid-2006. The number of NSDUH short reports dropped from thirty in 2005 to fifteen in 2006. Responses to data requests, including requests from ONDCP, took longer to complete or were not done at all. A survey of

OAS employees found only 14 percent satisfied with the policies and procedures of senior leaders.

OAS Recovers and Rebuilds

The near collapse of OAS was short-lived. Stephanie Colston left SAMHSA in October 2005. Administrator Curie left in 2006, and Deputy Administrator Dr. Eric Broderick essentially directed the agency starting months before Curie officially left. Alerted to the problem in OAS, Broderick took action to stabilize the staffing and rebuild OAS. He appointed me as a temporary associate director of OAS, reporting directly to him, to monitor the situation in OAS and prevent further loss of staff and damage to the OAS data systems. Then in October 2006, he replaced Kaiser with Dr. Anna Marsh as Acting Director. She was a longtime SAMHSA senior manager who had been Deputy Director of OAS under Donald Goldstone and also Deputy Director of CMHS.

It's unknown what would have happened to OAS and the NSDUH had Broderick not intervened. Undoubtedly more staff (including analysts and survey management personnel) would have left and analysis output would have declined further. Oversight of the contractor would have diminished, possibly resulting in cost increases and inefficiencies. But it is unlikely that cuts in the NSDUH questionnaire or sample would have been allowed by OMB and ONDCP. Evidence of this was seen in 2007, when we were asked by the SAMHSA budget office to find ways to reduce the cost of the project. We identified a few small potential savings in analysis, field management, and methodological studies, and explained that a substantial cut could only be achieved with a reduction in the sample size. But we cautioned that a large sample cut could possibly affect the data because it could alter the characteristics of the field staff. Research had shown that NSDUH respondents tended to report less drug use when interviews were conducted by interviewers that had worked on the NSDUH for many years. Fearing any further break in trend measurement, budget officials in HHS asked us to determine the maximum cut in the sample that would not impact trends. This was not something we could precisely determine, but we conservatively replied that a 5 percent sample reduction was very unlikely to affect the data, and would save about $1.5 million per year. SAMHSA submitted the budget to OMB with the 5 percent sample cut, and we instructed RTI to reduce the sample for the 2009 NSDUH from 67,500 to about 64,000. But ONDCP intervened, ordering HHS not to reduce the sample size. In January 2008, the funds were restored and the sample size was increased back to 67,500.

In 2006 the NOMS program was implemented, and states began to report data to SAMHSA. No major changes to the NSDUH questionnaire were required, and NSDUH estimates were used for some of the NOMS. I became an active participant in the development of the data strategy, along with a dozen other senior SAMHSA officials who met numerous times and drafted and reviewed several versions of the strategy. It was finally completed and published in 2007. Less than two years later, the data strategy essentially disappeared when the Obama administration began in 2009.

With SAMHSA (Eric Broderick) and OAS (Anna Marsh) leaders firmly supporting the survival of OAS and its programs, morale quickly improved.[2] By early 2007 four new highly skilled analysts had accepted positions in the NSDUH division.[3] Acting Director Marsh developed a reorganization plan, reducing OAS to just two divisions: Population Surveys (NSDUH) and Operations (all other data systems). There would be no resurrection of the Division of Analysis, but analytic capability would be strengthened in each of the two remaining divisions. Besides stabilizing and rebuilding OAS, a priority for Marsh during her brief stint as acting director was to enhance collaboration and reach out to other SAMHSA offices to offer the OAS data collection and analysis they needed to help manage their programs. She set up meetings with leaders from each center, in which we presented information on the NSDUH design and the capabilities of the data, including our ideas for studies we could do to help them in their work. We also told them about new possibilities for access to NSDUH data files (see discussion in the next section). Center staff requested particular analyses they were interested in, and suggested new items to be added to the questionnaire.

Sharing Data While Protecting Respondent Confidentiality

Managers of federal surveys have a responsibility to protect the confidentiality of the respondents. Protecting confidentiality is not only ethically and legally important, but also critical for the success of surveys like NSDUH that collect sensitive data that could result in serious consequences for a respondent if disclosed to others. The methodology used and the information provided to sampled individuals should convey to respondents that the survey contractor, the government officials directing the survey, and anyone authorized to access the data will never disclose their personal information. If citizens selected in the sample do not trust the survey organizations and staff, they may refuse to participate, or they may complete the interview but not report their drug use, causing bias in survey estimates. Each survey and organization has its own ways of addressing confidentiality, depending on the type of data collected, its intended uses, and the laws that pertain to them. NSDUH communications with potential respondents have always included strong promises that their individual information on drug use and other personal characteristics will never be disclosed and that the data will be used only for statistical purposes. Consistent with those promises, there are restrictions on public data releases. Data tables in reports do not show estimates for very small numbers of respondents. Descriptions of the sample do not reveal the specific geographic locations where interviews were done. Public use data files, which do include the actual data for individual survey respondents, are altered to prevent a clever "data snooper" from having the ability to identify any respondent. This is not trivial. With knowledge of just a few characteristics such as age, sex, race/ethnicity, state, and occupation, a snooper could identify some sample persons. Furthermore, we had to prevent the possibility of disclosure to an inside intruder, meaning someone who knows that a certain person was a survey participant. For example, the public use file could not allow a parent to find the data for their child who had participated in the survey. The result of this disclosure limitation was a reduced data file. Variables or combinations of variables that could reveal the identity of a respondent were removed. This process limited the kinds of studies that outside researchers could do. We were often approached by researchers requesting access to the full data set, or at least a few of the variables not included on the public file, such as state codes. The inability to share the full data set was frustrating for OAS, because we understood the value of the research they wanted to carry out. Despite extensive efforts over many years, working with HHS lawyers, we were not able to devise a satisfactory mechanism to make the full data set available to researchers. Until 2008.

In 2002, a new law impacting confidentiality in federal surveys was passed. The Confidential Information Protection and Statistical Efficiency Act (CIPSEA) contained provisions that were potentially beneficial to SAMHSA, OAS, and NSDUH in several ways. The law and the subsequent guidance issued by the Office of Management and Budget defined official federal statistical agencies and units as organizations of the executive branch, whose primary duties are the collection and analysis of data for statistical purposes.[4] Approved statistical units have certain responsibilities and authorities. Responsibilities include maintaining high standards of statistical integrity and objectivity and operating with independence from political influence in designing surveys and reporting results. The law requires that survey data be used for statistical purposes only, and that federal survey staff committing a confidentiality breach face penalties of up to five years in prison and a fine not to exceed $250,000. This provision gives survey organizations a tool to use in convincing sample individuals to participate in surveys, easing respondents' concerns about possible mishandling of their personal information. This may help surveys throughout the federal statistical system achieve high response rates. We were most excited about the authorities given to statistical units. Under CIPSEA, statistical units are allowed, at their discretion, to designate agents who can access the confidential data from the survey. The agents have to meet certain criteria, and are be subject to the same penalties if they violate confidentiality. Designation as an official statistical unit could enhance the prestige of SAMHSA and OAS, improve survey participation, and allow us to share state codes and other variables of interest with outside researchers. We immediately changed the introductory materials on the survey, mentioning CIPSEA to potential respondents beginning with the 2004 survey.[5] Then in 2006 we prepared materials and a justification to become a statistical unit. A few days after Anna Marsh became acting OAS Director, we submitted the request to SAMHSA leadership. There were some concerns about the implications, but after we explained the benefits of OAS becoming a statistical unit, with HHS legal counsel providing support, SAMHSA agreed to move ahead with the request. Acting Deputy Administrator Eric Broderick sent the request to OMB Chief Statistician Katherine Wallman. Three weeks later, on November 9, 2006, OMB approved the request and OAS was officially a federal statistical unit.

Armed with our new legal authority, survey methodologist and confidentiality expert Jonaki Bose, with help from mathematical statisticians Art Hughes and Jim Colliver and a team at RTI, led a successful effort to develop new protocols to allow access to NSDUH restricted-use data files. The program was piloted in April 2008, initially making files available to SAMHSA employees and their contractors, and expanded to all outside researchers in 2012 through a data portal. Researchers develop proposals and submit them to CBHSQ. If CBHSQ determines the study to be feasible and data security to be adequate, the study is approved and the researchers are given access to the restricted use files remotely through a portal. Output is monitored by CBHSQ to ensure there is no disclosure risk, but researchers get access to the restricted-use data and software capable of various statistical analysis methods. The portal also can create linkages of NSDUH data to external data sources, or create customized subfiles and recodes as desired.

Bose and her team also implemented a Restricted Use Data Analysis System (R-DAS), which allows researchers to conduct basic analysis of the restricted files online, without actually downloading the confidential files. Output is controlled by software that is designed to prevent analyses of combinations of variables that could disclose information on individual respondents. The files accessed through R-DAS contain state and county codes as well as detailed demographic data on each respondent.

OAS Becomes a Center

Communication difficulties between the survey staff and senior SAMHSA management began to recur after a new OAS director, Dr. Peter Delany, was appointed in January 2008. Trained in social work, he did not have a strong background in survey research or statistics. OAS statisticians struggled to explain and get him to accept basic principles of survey research such as the importance of using weights, providing complete information on methods and sampling errors in reports, the use of sampling errors and statistical significance testing in interpreting results, and the procedures required to implement valid questionnaires and data collection methods. The limited understanding of survey research by Delany and some of his senior managers, and their reluctance to consider statistical and other concerns raised by survey staff, led to inefficiencies, mistrust, and animosity. Communication between the senior managers and survey teams was poor. Problems worsened after a SAMHSA reorganization expanded the scope of OAS, and the office became the Center for Behavioral Health Statistics and Quality (CBHSQ) in late 2010. The new center would have more than twice the number of staff that OAS had, including a large new evaluation and analysis division.[6] Inquiries about NSDUH data, requests for special analysis of NSDUH data, and some NSDUH questionnaire development work was handled by staff in the new division who had little knowledge of the survey or survey design in general.

These communication and management problems led to another substantial drop in morale, and loss of staff.[7] Nevertheless, thanks to the resilience of the remaining staff and their dedication to the project and to statistical integrity, there was no major damage to the survey. Eventually, in 2015, there were management changes that resulted in a more congenial and constructive working environment.[8] The CBHSQ statisticians were successful in protecting the core data collection and maintaining valid trends until at least 2014. In part because CBHSQ was an OMB-approved statistical unit, we had some success in preventing leadership from eroding statistical standards and best practices for federal surveys.

Annual Releases of 2003–2014 NSDUH Results

The annual release of the survey results at the Recovery Month kickoff event became routine after the controversial 2002 report that established a new baseline. Each year, the event took place on the first Thursday after Labor Day at the National Press Club in Washington, DC. From 2006 to 2013, the event was moderated by CSAT Director Dr. Westley Clark, with speeches given by the SAMHSA administrator and the drug czar,[9] followed by brief talks by individuals in recovery. The stories they told were often emotional and showed courage and appreciation for the help they received in their battle with drug addiction. The event became a celebration of those who had achieved recovery, a recognition of treatment providers, and a notice to persons with behavioral health problems that help is available and people do recover. The NSDUH data described the scope of the problem and the latest trends, providing a reminder to policymakers that continued support for programs was needed to address the problem. During these years, increases in problems associated with nonmedical prescription opioid use, heroin use, and marijuana use were often highlighted. The use of cocaine, cigarettes, and alcohol declined during this period.

Political controversies surrounding the survey results rarely happened during this period, in part because the release became somewhat routine and was consistently scheduled as part of the Recovery Month kickoff day. However, in 2012, there was a delay in the release of the 2011 results. The press conference was initially set for September 6, but that turned out to be the day President Barack Obama was speaking at the Democratic National Convention. The Recovery Month kickoff and NSDUH release was pushed to September 24.

Within HHS, there were differences of opinion about whether the event should be limited to substance abuse recovery and data, or if mental health should also be included. After the new mental health questions were added to NSDUH in 2008, SAMHSA decided to publish separate annual National Findings reports for substance use (released at the Recovery Month kickoff) and for mental health (released a few months later). This essentially kept the September event limited to substance use issues, consistent with the annual presidential proclamation designating September as "National Alcohol and Drug Addiction Recovery Month." But SAMHSA Administrator Pam Hyde wanted the event to focus more broadly on recovery from both mental illness and substance use disorders. Beginning in 2011 the event was referred to by SAMHSA as simply "Recovery Month." Materials and speeches included mental health concerns more prominently.

The CMHS director participated in the event. However, the NSDUH report and data release at the kickoff event continued to include only substance use data. It was at times awkward, such as in 2012 and 2013 when Hyde's speech at the press conference cited old NSDUH statistics on mental illness along with the new substance use results. The mental health data were from the prior year and were not in the National Findings report released at the press conference. In early 2014 Hyde informed CBHSQ Director Pete Delany that for the 2013 data release, which would occur in September 2014, the NSDUH report must include both mental health and substance use results. Unfortunately, Delany neglected to tell the NSDUH team of this important policy shift, so data production proceeded as it had the year before: A National Findings report on substance use would be ready for the September press conference. When Administrator Hyde saw the finished report with no mental health data, she was angry with Delany. NSDUH staff were instructed to quickly pull together a short report with some key findings on both substance use and mental health, and have it ready for the press conference. The following year, the data processing and reporting schedule for the 2014 survey was revised to make sure there would be a full release of the 2014 mental health data along with the 2014 substance use data, at the Recovery Month kickoff press conference in September 2015. However, this time the report format was significantly different. The comprehensive reports published as Main Findings and National Findings for over forty years were discontinued. Instead, a brief summary report covering selected trends in substance use and mental illness was prepared,[10] supplemented with a series of other short reports covering specific topics (risk and protective factors, initiation, services received, suicide thoughts and behaviors, and teen depression) released in the subsequent weeks.

A recurring conundrum associated with the NSDUH releases at the Recovery Month events was that NSDUH did not produce any estimate of how many Americans were in recovery. A question about this came up at several of the press conferences, and each time SAMHSA officials replied that we were working on it and would have it soon. For at least a decade, various workgroups were convened in SAMHSA to try to develop a measurable definition of recovery. Finally, after soliciting public comment, in March 2012, SAMHSA released its official definition of recovery from mental disorders and/or substance use disorders. Recovery was "A process of change through which individuals improve their health and wellness, live a self-directed life, and strive to reach their full potential." As a concept, it was well received by recovery advocates. But to statisticians in CBHSQ, it was not clear that this definition could be reliably assessed and measured by a small set of survey questions, and little progress was made toward adding questions to the survey.

Estimating First-Time Drug Use (Initiation)

The limited capability to compare 2002 and later data to 2001 and earlier estimates led SAMHSA to rely more on retrospective incidence estimates[11] to describe long-term trends. For example, the 2002 National Findings report shows the annual number of new nonmedical pain reliever users increasing from 628,000 in 1990 to 2.4 million in 2002.[12] These estimates, constructed from the reported age at first misuse of pain relievers among the 2002 NSDUH respondents, contributed to widespread concerns about a rise in opioid use in the early 2000s.[13] But a methodological problem with these data was identified in 2004. Recall bias refers to inaccurate reports of prior events due to memory failures. It is a well-known occurrence in surveys, and the severity typically depends on many factors, including the salience of the event being asked about and the length of time since it occurred. Not surprisingly, it turns out that respondents are better able (or willing) to accurately report their initial use of a drug if it occurred recently than if it occurred many years ago. Another factor that seems to affect NSDUH data on age at first use is telescoping. Telescoping refers to reporting use of a drug, but recalling that the age at first use was more recent than when it actually occurred. These two types of reporting errors combine to cause NSDUH retrospective incidence estimates for past years to be underestimates. The underestimation increases as the length of time between the survey year and the year of the estimate increases, and varies by drug.[14] Figure 11.1 shows that the bias appears to be much worse for nonmedical use of pain relievers than for other substances.

The estimate of pain reliever initiation in the year 2003 is 1.45 million based on interviews done in the next year (2004), but only 0.83 million based on interviews done eleven years after the initiation occurred (2014). This pattern of estimation bias can create the appearance of an increasing incidence trend over the long term, when in reality there may be no change at all. Because of the discovery of this bias inherent in the retrospective estimation method, we decided to discontinue use of the method for long-term trends. With the 2004 report, we began a new series of incidence measures, which describe initiation in the twelve months prior to interview. Not only did this limit the recall period to twelve months, it also provided consistency in the reference period for several key NSDUH measures such as past twelve-month use and substance use disorder.

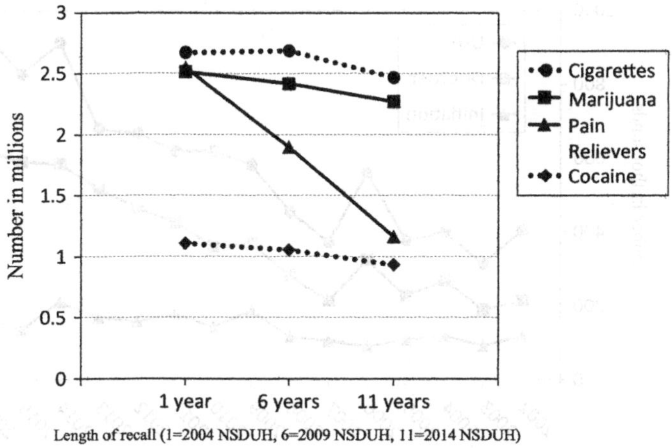

Figure 11.1 Estimated number of substance use initiates during 2003, by length of recall.

A Powerful Analytic Tool

NSDUH's consistent methodology from 2002 through 2014 resulted in a data set with tremendous potential for the study of substance use in the United States. The country paid more than a half billion dollars to compile these data, collected from more than 900,000 respondents, including more than 11,000 in every state. Unlike many large national samples that are concentrated in a few hundred large geographic clusters, such as groups of counties, this sample covered virtually all areas of the United States. The thirteen-year sample covered more than 90 percent of US counties, including about 400 counties each having representative samples of more than 500 respondents. In addition to the annual National Findings reports, OAS and CBHSQ produced many other studies, such as:

- Reports with tables and maps of model-based state estimates for about twenty-five substance use indicators were produced each year, a few months after the initial releases of the National Findings. These estimates were based on pooled data from the two most recent surveys. For example, the report on the 2004 survey contained estimates using the combined 2003–4 data. Eventually trends by state were assessed in the reports. Every two years, a substate report was produced, using pooled data from the most recent three surveys. These reports provided estimates for the same twenty-five or so measures, for over

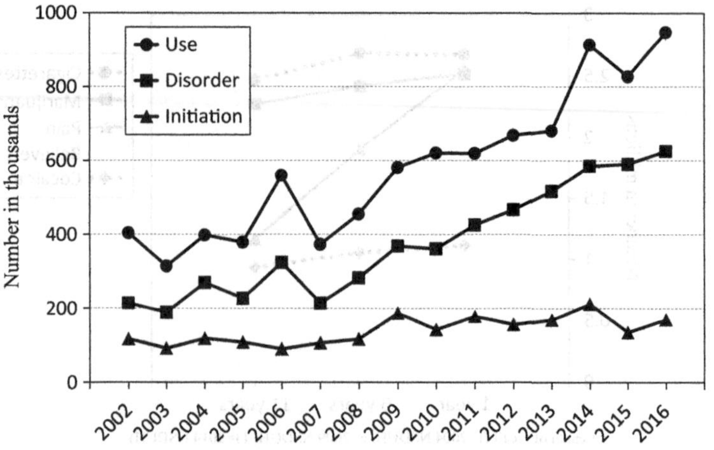

Figure 11.2 Heroin initiation, use, and disorder (abuse or dependence) in the past twelve months, by year.

300 substate areas that CBHSQ had defined in consultation with state substance abuse analysts. The areas were defined based on the needs of state officials, with the constraint that the sample size had to be sufficient to support reliable estimates based on three years of data.

- Studies of opioid abuse showed a decline from 2002 to 2013 in nonmedical use of prescription pain relievers, but an increase in pain reliever dependence and abuse and a link between prescription pain reliever abuse and heroin use. 80 percent of heroin initiates had previously used prescription pain relievers nonmedically.[15] Despite the assumed undercounting of heroin abuse by NSDUH, the survey has been consistent and an important source of data tracking the recent heroin epidemic using different indicators, shown in Figure 11.2.
- Several studies looked at the patterns and trends in drug use among older adults. The projection model developed in an earlier study worked well with older adults and showed the number of persons age 50 and older needing substance abuse treatment would more than double between 2004 and 2020 as the baby boomers grow older.[16]
- Combined 2002–07 data provided a large enough sample to study the patterns of use among women prior to pregnancy, during pregnancy, and postpartum. The study found significant reductions in alcohol and marijuana use during pregnancy, but resumption of use after giving birth. 17 percent of pregnant women smoked cigarettes.[17]

- The highest rate of initiation of marijuana use among teens was found to occur during the summer months. ONDCP used NSDUH data based on month of first use questions added in 1999 to warn parents ("ONDCP Alerts Parents: Marijuana Use Spikes When School's Out") about the increased risk during summer.[18]
- Analysis of trends during the period when many states were legalizing medical and other marijuana use found a major decline in the percentage of Americans believing there was a great risk of harm in using marijuana once or twice a week, from 53 percent in 2003 to 34 percent in 2014. This trend coincided with increases in initiation among young adults age 18–25 (from 700,000 to 1.1 million) and adults age 26 and older (from 100,000 to 300,000), but not youths (1.2 million in 2003 and 2014).[19]
- Detailed comparisons of NSDUH estimates with estimates from other surveys identified the methodological factors that may explain differences in prevalence estimates for youth surveyed in school versus home, and different estimates of treatment based on population survey reporting versus administrative data compiled by treatment facilities.[20]
- Taking advantage of the geographic coverage and continuous data collection of the NSDUH sample, studies assessed changes in substance use and mental health after specific events such as Hurricanes Katrina and Rita, which struck the Gulf Coast in August and September 2005;[21] the economic downturn of 2008;[22] and the Deepwater Horizon oil spill in 2010.[23]
- Studies of the impact of the Patient Protection and Affordable Care Act of 2010 showed estimates of substance abuse and mental illness in the populations that would be newly covered by Medicaid.[24]

Considering all of the studies summarized, other research studies not mentioned, all of the annual reports on national, state, and substate findings, and other SAMHSA reports (one-page spotlights, four- to twelve- page short reports, and longer data reviews), SAMHSA funds for NSDUH analysis amounted to roughly $100–150 million between 2002 and 2015. This is less than a quarter of the total cost of the surveys. Recognizing the limited budget and staffing for analysis by the project staff, NIDA, OAS, and CBHSQ have placed a high priority on providing public use data files to outside researchers. Recently, files have been available within a year after the completion of data collection. A file is available for every survey since 1979. SAMSHA also developed new tools for researchers to obtain access to restricted-use NSDUH files.

The expanded sample and new tools for access and analysis led to a substantial increase in the use of the data by researchers outside SAMHSA and its contractors. The annual number of peer-reviewed journal articles, reports, and dissertations based on analysis of NSDUH data and not supported directly by OAS or CBHSQ increased modestly from thirty-six in 2000 to sixty-one in 2010, then jumped to 123 in 2014.

Notes

1 In survey terminology, "precision" typically refers to sampling error, which is highly dependent on sample size, but has little to do with the length of the questionnaire.

2 The government-wide Employee Viewpoint Survey result for the percent of OAS staff satisfied with policies and procedures of senior leaders was 14 percent in 2006 and 74 percent in 2007.

3 Beth Han, Jonaki Bose, Jim Colliver, and Lisa Colpe.

4 "Statistical purpose" is defined in CIPSEA as the description, estimation, or analysis of the characteristics of groups, without identifying the individuals or organizations that comprise such groups.

5 OMB allowed us to invoke CIPSEA even though OAS was not an official statistical unit, as the process for becoming a statistical unit was not yet developed. CIPSEA designated an initial set of statistical agencies and units, but OAS was not included. In 2006, we were advised by OMB to submit a request for designation as a statistical unit.

6 The funds to cover staff increases were made available by the closing of the DAWN survey.

7 The percent of OAS staff satisfied with policies and procedures of senior leaders dropped to 50 percent in 2010 and 18 percent in 2015.

8 In November 2015, Delany was detailed to a new position at ONDCP. Daryl Kade was appointed Director of CBHSQ.

9 SAMHSA administrators and press conferences attended (by year of event) were Charles Curie (2002–05), Ric Broderick (2006 and 2008–09), Terry Cline (2007), Pam Hyde (2010–14), and Kana Enomoto (2015–16). Drug Czars were John Walters (2002–08), Gil Kerlikowske (2009–12), and Michael Botocelli (2013–16).

10 CBHSQ, *Behavioral Health Trends.*

11 Retrospective incidence estimates indicate the number of persons using a drug for the first time during a defined period of time, such as a calendar year. The estimates are constructed from respondents' reports of their age at first use of a drug and date of birth. For example, a 40-year-old respondent in 2004 who reports he was age 20 when he first used marijuana is counted as an initiate in 1984.

12 SAMHSA, *Results From the 2002 NSDUH*, 46–7.

13 Compton and Volkow, "Major Increases in Opioid," 103–7.

14 SAMHSA, *Results from the 2004 NSDUH*, 45 and 121–27; Gfroerer et al., "Estimating Trends in Substance Use."

15 Muhuri, Gfroerer, and Davies, *Associations of Nonmedical Pain Reliever Use*; Han, Compton, Jones, and Cai, "Nonmedical Prescription Opioid Use," 1468–78.
16 Han, Gfroerer, Colliver, and Penne, "Substance Use Disorder," 88–96.
17 Muhuri and Gfroerer, "Substance Use Among Women," 376–85.
18 SAMHSA, *The NSDUH Report: Seasonality*.
19 Lipari, Kroutil, and Pemberton, *Risk and Protective Factors*.
20 SAMHSA, *Comparing and Evaluating Youth Substance Use*; Batts et al., *Comparing and Evaluating Substance Use Treatment Utilization*.
21 SAMHSA, *The NSDUH Report: Impact of Hurricanes Katrina and Rita*.
22 Compton, Gfroerer, Conway, and Finger, "Unemployment and Substance Outcomes," 350–3.
23 Teich and Pemberton, "Epidemiologic Studies of Behavioral Health," 77–85.
24 SAMHSA, *The NSDUH Report: Trends in Insurance Coverage*.

12 How to Redesign an Ongoing Survey, Or Not

After the survey changes in 2002 resulted in a break in trend measurement, a primary concern in managing the survey was to maintain comparability with 2002 data, the new baseline. Any proposed change to the questionnaire or other aspect of the design and estimation was scrutinized by the survey team for the potential to cause estimates to change. We were able to maintain consistent data for most key substance use indicators for at least fourteen years. Yet during that period we knew that updates would have to be made to the survey at some point as data needs shifted. There were emerging concerns about methamphetamine and opioid abuse, drug abuse among aging baby boomers, and policy changes such as marijuana legalization and health care reform. Updates to implement improved data collection methods might also be warranted. We decided we needed a long-term plan for managing the survey design that would guide future decision-making and provide a framework that would prevent disruptions in trend measurement and also support periodic redesigns as needed.

Conceptual Approval for a Redesign

We outlined a proposed survey design management plan in early 2005. The plan stated that every ten years there would be a major redesign of the survey methods and questionnaire. The redesign would be implemented, if possible, in a way that provides a bridge or adjustment method for long-term measurement of trends for the most critical variables of interest. The next redesign would be in 2012. There were several advantages to this approach. First, having a set date for redesign would facilitate planning a series of methodological studies, making sure the necessary research, testing, and development of new methods (including questionnaire content changes) were ready in time for the 2012 implementation. Second, having a set date for redesign would help us resist pressure to change the questionnaire every year. Significant changes would be made during interim years only if major methodological problems emerged, or if critical new policy issues needed to be addressed immediately. Finally, official approval of the plan would help ensure that the redesign would actually occur. Without such a structured plan and organizational commitment, those most concerned about trend measurement would resist any attempt to make methodological improvements or content updates to the NSDUH.

I presented the survey design management plan to OAS Acting Director Dr. Charlene Lewis and the NSDUH Expert Consultant Panel in April 2005. There was unanimous agreement that SAMHSA should proceed with this approach. After the meeting, we began developing the details of the plan. We needed a methodological research program, a process for obtaining input from data users, and a timetable for field testing and implementation of the new design. We also needed to announce the plan to other government agencies and to the public, and have an open and transparent process that would involve constituents and data users. To that effect, I inserted a section "Planning for a Redesign of NSDUH and Trend Measurement" in the final discussion chapter of the 2004 National Findings report. The section ended with the statement "Having a systematic plan that preserves trends and allows for survey improvements will help maintain project continuity and consistency at times when there are changes in administrations, agency leadership, and project management."

Unfortunately, communication between OAS and SAMHSA leadership was lacking during this period of transition for OAS. I should have recognized that under the circumstances our plan and its rationale may not have been explained to or vetted with SAMHSA leadership and ONDCP. The National Findings report had been approved by SAMHSA and HHS officials, with no concerns raised about the future redesign described in the report. Copies were printed and ready to be distributed at the September 8 press conference. But in August when I briefed SAMHSA Administrator Curie and his senior staff on the results of the survey, I mentioned the ten-year redesign plan. They were unaware of it. Curie was furious. He knew how upset ONDCP Director John Walters was about the break in trend when we changed the name of the survey and began the thirty dollar incentive payments two years before. Any *mention* of another change to the survey that could disrupt trends was forbidden. There could be no evidence that SAMHSA was considering changing the survey again – even if it was to occur seven years in the future. Curie ordered all printed copies of the report be destroyed. Not only would they be destroyed, but the offending page first had to be torn out of each report and shredded, so that there was no possibility that any evidence could be leaked and discovered by ONDCP. Newly hired OAS Director Javid Kaiser rounded up all available OAS staff to spend a day tearing out the forbidden pages.

It was ironic that the plan designed to prevent unplanned breaks in trends was extinguished because of concerns it would lead to a break in trend. Even more ironic were the vigorous efforts by Kaiser and Curie advisor Stephanie Colston a few months later to overhaul the questionnaire, which probably would have caused a break in trend.

A year later, an analysis of costs and budget showed an approaching shortfall in funding for NSDUH. A reduction in costs was needed. By that time Curie had left SAMHSA. I suggested to Acting Administrator Eric Broderick that the previously squashed survey redesign plan could be revived with an added goal of reducing NSDUH costs to meet anticipated budget levels. In developing a new design, we could identify less costly approaches for data collection, processing, and analysis, and ultimately end up with an optimal design under new cost constraints. Broderick liked the idea, so we resumed redesign work in early 2007.

Areas for Improvement

By 2007, a variety of potential design changes to improve the survey had been identified. For example, the need for more data on issues such as mental and physical health, health care utilization, and older adult drug use suggested that larger samples of adults would be useful. The redesign in 1999 allocated only one third of the sample (22,500) to adults age 26 and older – an age group that constituted 77 percent of the population. Without reducing the overall sample size of 67,500, a reallocation of the sample from the younger groups to the older group would reduce the number of addresses needed to be contacted and screened, lowering costs.[1] Similarly, adjustments to state sample allocations could be considered to improve the precision of state and national estimates.

Potential improvements in the NSDUH questionnaire were identified. Several items asked in the interviewer-administered sections were considered candidates for inclusion in the ACASI to improve data quality. These included questions about income, skipping work or school, and other sensitive topics that would best be answered in private. Refinements to enhance question clarity or improve the flow of questions were proposed. The prescription drug modules needed restructuring. The set of drugs asked about in the core modules was out of date. Several new drugs that had not been added to the core modules were frequently reported in the "other-specify" questions, and studies have found that direct questions specifically naming the drugs result in more accurate reporting.[2] Methamphetamine questions embedded within the prescription stimulant module needed to be moved into a new stand-alone methamphetamine module, acknowledging it as a street drug typically manufactured in clandestine labs. Also, there was a need for more detailed information about the misuse of specific drugs in the past twelve months, instead of in the lifetime.

Methodological Research to Guide Design Decisions

In the fall of 2007, the NSDUH team held internal meetings and worked with RTI to compile existing methodological research studies that could inform the redesign, and determine what new methodological studies were needed. Several priority areas of investigation were identified: sample design, improving response rates, questionnaire, and estimation. Some issues identified for investigation were assessing alternative sample designs and coverage; evaluating the use of available address listings for sampling (instead of having field staff create listings); developing updated and improved contact materials (such as lead letters); testing alternative question orders to improve response accuracy; overhauling the prescription drug modules; converting photographs of prescription pills (pill cards) from hard copy to electronic form; and evaluating alternative ways of editing, imputation, and weighting.

Assessing Data Users' Needs

A critical task, begun in mid-2008, was to conduct comprehensive outreach to key NSDUH data users (SAMHSA programs, other federal agencies, states, outside researchers), to obtain feedback on their data needs, and to solicit specific recommendations on the redesign of NSDUH, especially the questionnaire content. Feedback from these contacts would guide methodological development and finalization of the new survey design. An important component of the outreach effort was a web-based survey on state governmental data needs, conducted in May 2008. The survey asked representatives from every state whether or not their state used the NSDUH data. All fifty states reported being aware of the NSDUH data and forty-nine states reported using the data. Thirty-seven states reported using NSDUH data for developing policy and legislation, and twenty-two states reported using the data for allocating funds. Based on these results, the redesign work proceeded under the assumption that state estimation capability would be retained.

Advice from Outside Experts

After the plan for methods studies was developed, we convened our expert consultant panel. The meeting took place on November 6, 2008. This was two days after Barack Obama was elected president, so the implications of that were discussed at the meeting. We were uncertain of the impact of the election result, except there would be new leadership at ONDCP, HHS, and SAMHSA.[3] They would need to be briefed on the redesign, and they could tell us to change the plan. I wondered if Vice President Biden's involvement with the survey data two decades earlier would have any influence. As chair of the Senate Judiciary Committee, his comment on the 1990 survey results, when reported by the Bush administration, was that the survey was "wildly off the mark."[4]

The entire expert panel meeting focused on the redesign. We discussed our proposed methodological studies, and asked the panel if our overall plan was reasonable and what changes in survey design and questionnaire content should be made. Most panel members thought the plan was good, and they were impressed with the thoroughness of the approach. Several consultants said maintaining trend capability was more important than making improvements, but if many changes were going to be made, they should be made simultaneously and SAMSHA should "bite the bullet" on trend measurement for that point in time. There was skepticism as to whether a dual sample approach to "bridge" the before and after samples would succeed. There was also a suggestion by Dr. Groves that SAMHSA consider a flexible design, allowing for frequent changes to the design to meet emerging data needs as they occur, instead of waiting for the redesign every ten years. He also urged SAMHSA to consider a mixed-mode design, in which some data might be collected by phone, mail, or internet mode, in addition to the face-to-face computer-assisted mode.

Managing the Redesign Work While Continuing the Survey

A challenge faced in managing the redesign effort was staffing. SAMHSA did not have a designated survey design unit that could develop the design while the NSDUH team continued to manage the regular activities associated with data collection and analysis. Of course we relied heavily on our contractor (RTI), but ultimately the staff in my branch made enormous efforts to manage, direct, and develop the redesign while the regular survey work went on. They had to oversee the extra redesign work, planning studies, reviewing complex contractor reports, and making decisions based on in-depth discussions of results and their implications. We held regular meetings, referred to as YARMs (Yet Another Redesign Meeting), to discuss progress. The burden on staff, especially the Instrumentation Team, was stressful, particularly given the uncertainties of the outcome. A great deal of effort went into development of questionnaire improvements that we expected to cause a break in trend, but there was no guarantee that ONDCP and SAMHSA leadership would allow that result.

We pushed the implementation target year back to 2015, partly to alleviate the pressure on staff, but also because more time was needed to finish some of the methods studies and testing of new instruments. We also hired new staff. New project officers Michael Jones and Dr. Peter Tice were recruited to relieve Art Hughes, who had been project officer since 2000 and was heavily involved in redesign activities and state estimates.[5]

When federal agencies decide to redesign a large ongoing survey, they often employ a separate contractor, and sometimes separate government teams within the program, to develop the design. With this approach, the staff (federal and contractor) with responsibility for managing the day-to-day operations can be minimally affected. But there are advantages to integrating the design work within the main survey activity, as we did with the NSDUH redesign. Staff with in-depth knowledge of the survey goals, design, operations, and data uses can best identify areas for improvement and feasible approaches to evaluate them. Often, as was the case with the NSDUH redesign, methodological studies associated with a redesign can be embedded within the main survey activities, saving costs and providing realistic comparisons between old and new methods.

Final Design Decisions

By 2010 we had completed enough outreach and research to be able to narrow down the options for the new design. To complete the remaining work to implement the redesign in 2015, a critical policy decision was needed. The choice was between a full redesign incorporating all identified improvements with a break in trend, or minor changes and updates in the non-core sections of the interview to preserve trend capability as much as possible. SAMHSA and ONDCP leadership would have to make this decision. Although those of us who had been immersed in the redesign work for the past three years were convinced we were prepared to substantially improve the survey with the proposed design changes we had developed, I sensed that ONDCP and SAMHSA would not allow a break in trend. There was not enough evidence that the current design was all that bad. So I decided to add a third, middle-ground option, in which a few critical design improvements were made, but some of the core modules would be retained, to limit the impact on trend measurement. Under this option, the prescription drug modules would be completely redesigned, the new methamphetamine module would be created, and the contact materials would be updated. There would be no restructuring and only minor rewording of the initial core drug modules on tobacco, alcohol, marijuana, cocaine, heroin, inhalants and hallucinogens, which all preceded the prescription drug modules in the interview flow.

On April 30, 2010, the NSDUH team met in the OAS conference room with Terry Zobeck, ONDCP's Associate Director of the Office of Research and Data Analysis, to discuss the three options.[6] As expected, he was averse to any redesign that would change estimates and disrupt tracking of major drug abuse trends. We were careful to point out that although the partial redesign option was highly unlikely to cause changes to the estimates for major illicit drugs of abuse, there was a slight chance there could be a small impact, especially for inhalants and hallucinogens. Also, the prescription drug module changes would definitely change results for nonmedical use of those drugs, and disrupt trend measurement for indicators such as any illicit drug use and illicit drug treatment need. Nevertheless, recognizing the need for some improvements, Zobeck chose the partial redesign approach. In accordance with the ONDCP preference, OAS Director Pete Delany decided that much of the work on the redesign should stop. Only activity that was related to the partial redesign plan would continue.

The NSDUH team was disappointed that we would not be implementing the full redesign, but relieved that *some* of the improvements we developed could be implemented, and the remaining design work could be more focused. However, we still needed approval from SAMHSA Administrator Pam Hyde. That would end up taking more than a year. During that time, relevant methodological development proceeded and details of the plan were hashed out under the assumption that the partial redesign plan would be approved.

I never saw a final, formal approval of the redesign plan from Pam Hyde or the Director of ONDCP (Gil Kerlikowske). I assumed Zobeck discussed the redesign with Kerlikowske. Delany said he discussed it with Hyde. I was concerned Hyde might not fully understand the rationale behind the redesign plan based on Delany's explanation, because he had not been involved in much of the redesign work, and he had limited knowledge of survey research methods and terminology. Delany asked me for a short summary document for his use in explaining the redesign plan to Hyde, but he did not ask me to participate in a discussion with her. Nevertheless, we moved ahead under the assumption that this plan would be implemented in 2015. The only official approval that I saw was an email to me from Delany in May 2011, in which he stated "Go ahead with the plan."

We conducted two field tests prior to the 2015 instrumentation redesign. The first one was in 2012 with a nationally representative sample of about 2,000 respondents interviewed during September through November. This test was designed to study the impact of the new instrumentation on data quality, length of the interview, and reporting of substance use.[7] The field test included a test of a new tablet device for household screening and case management. A final dress rehearsal field test with a nationally representative sample of about 2,000 respondents was done in September and October 2013, to further assess the questionnaire changes, and to test the Spanish version of the interview and a new lightweight laptop computer for interviewing.[8]

The partial redesign was implemented in two stages, summarized below[9]:

Sample design in 2014:

- State sample sizes were altered so the sample size in each state was closer to proportional to the state population.
- Age group allocations were shifted from 33 percent to 25 percent for age 12–17, from 33 percent to 25 percent for age 18–25, and from 33 percent to 50 percent for age 26 and older, closer to proportional to the age distribution in the population.
- The sample size remained at about 67,500, but the number of segments was reduced from 7,200 to 6,000, with increased within-segment sample sizes in large states.
- Field listing of addresses in sample segments was retained, with no use of existing address lists.

Survey materials and questionnaire in 2015:

- ACASI questions changed from pre-recorded human voice to computer-generated voice
- Contact materials were updated to reflect current organization names and to improve layout and visual appeal.
- Snuff and chewing tobacco questions were combined into single smokeless tobacco questions.
- The threshold for binge and heavy alcohol use for female respondents was changed from five or more drinks to four or more drinks on an occasion, to be consistent with official federal guidelines.

- Inhalants and hallucinogens lists were updated.
- A new methamphetamine module was added, separate from questions about misuse of prescription stimulants.
- The approach and definition for measuring the misuse of prescription drugs were revised, and questions about any use of prescription drugs (including medical use) were added.
- Questions about specific prescription drugs were changed from lifetime to past twelve-month misuse.
- Electronic images of prescription drugs replaced the hard copy pill cards that were shown to respondents, and examples other than pills were shown. The list of drugs was updated.
- The marijuana market module was removed.
- Demographic questions about moves in the past twelvemonths, immigrant status, marital status,[10] education, and employment were moved from being interviewer-administered through CAPI to being self-administered through ACASI.
- Questions for adults about sexual identity and attraction were added to the ACASI section of the questionnaire. Questions about disability status and how well the respondent speaks English, that HHS required all of its major surveys to include, were also added to the ACASI section.
- Questions about immediate family members currently serving in the US military were added.
- Industry and occupation questions from the employment section were removed. These were interviewer-administered questions that required extensive probing and interview time.

Redesign Effects on Estimates

CBHSQ and RTI analysts assessed the effects of the 2014 sample design changes and concluded they had no measurable impact on prevalence estimates.[11] Comparisons with 2013 and earlier estimates were included in the reports on the 2014 survey results.

The 2015 design changes, however, were expected to result in important changes in data that would restrict comparability. Assessment of the impact of the redesign on trends began before the 2015 data collection was completed, using data from the first half of 2015. Some unexpected results appeared. Further analysis with the full 2015 data began as soon as the data set was available in early 2016, confirming the preliminary findings. There were apparent trend breaks that had not been anticipated, and did not have definitive explanations. As expected, analysis showed that data on cigarette, alcohol, marijuana, cocaine, and heroin use were unaffected by the redesign. For these data, comparisons between 2015 estimates and estimates from 2014 and earlier years were valid. The analysis also showed that where questionnaire changes were made, estimates were affected, resulting in trend breaks. Thus, the 2015 data release did not include comparisons with estimates from prior years for:

- All smokeless tobacco estimates.
- Estimates of binge and heavy alcohol use for females, binge and heavy alcohol use for the overall population (both genders), and initiation of binge alcohol use.
- Estimates of overall use of any hallucinogen, past year initiation of hallucinogen use, and hallucinogen dependence and abuse.
- Estimates of overall use of any inhalant, past year initiation of inhalant use, and inhalant dependence and abuse.
- All estimates of methamphetamine use and past year initiation of methamphetamine.
- All prescription drug measures.
- All measures of any illicit drug use and the use of any illicit drug other than marijuana, and any illicit drug dependence and abuse (due to the changes to questions about hallucinogens, inhalants, methamphetamine, and prescription drugs).

There were other possible breaks in trends that had not been anticipated. Although the questions on substance use treatment did not change in 2015, and there was no clear evidence of a trend break for overall estimates of past year treatment, SAMHSA decided not to present trends for treatment data. There were concerns that some subgroup estimates may have been affected by questionnaire changes in earlier sections that impact how respondents are routed into the treatment module. Further analysis using subsequent years of data were planned, to more definitively determine if there was a trend break. Most surprisingly, a large shift in rates of perceived great risk was seen in the 2015 data (see Figure 12.1). NSDUH analysts were not able to conclusively determine what survey change could have caused this reporting change, although some form of context effect was surmised as the culprit. In the 2015 reports and tables, no trends were shown for drug treatment or perceived risk. Later, 2016 data suggested that the trends in perceived risk that had been evident in the data before 2015 were continuing (no change for cocaine, decreasing for marijuana), but at different levels.

Anticipating the upcoming release of the 2015 findings and the potential for criticism and misunderstandings regarding the reporting of trends, the NSDUH team wisely decided to publish a "heads-up" report

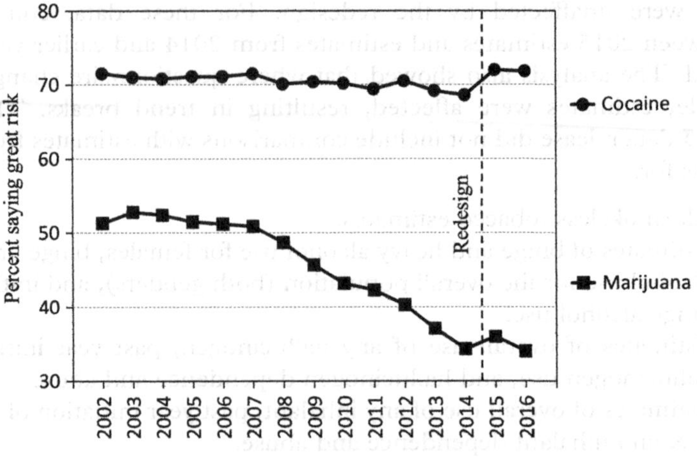

Figure 12.1 Perceived great risk in using marijuana once or twice a week and in using cocaine once a month, by year: ages 12 and older.

prior to the September data release. The brief report, published in July 2016, explained the survey changes, why they were made, and the effects on the 2015 data and trend capability. Although no data were included in the report, the results of the analysis and the specific measures that would not have trends reported, were listed.[12] The 2015 release went relatively smoothly, with few complaints or questioning about the loss of some trends. A comprehensive report on the impact of the redesign, with an extensive set of tables, was eventually published in June 2017.[13]

Other Changes Considered for the Survey

Although the major substance use indicators remained consistent throughout 2002–14, and the long-term plan for redesigning the survey every ten years or so had been approved in 2007, ideas for collecting new data persisted. In each case, the NSDUH team considered whether there were existing questions successfully used in other surveys that could be inserted into the NSDUH questionnaire with minimal developmental effort to provide reliable data without adversely affecting data currently collected, or if new questions needed to be developed. Decisions about adding or developing new questions were based on how important the new data were, whether the potential trend breaks due to context effects would be acceptable, how much interview time would be needed, and available time and resources. Another consideration was whether or not the goals of the proposed additions could be achieved through NSDUH, or if there was another survey that would be a better vehicle.

In 2007, ONDCP asked SAMHSA to explore two issues for potential inclusion in NSDUH. These were (1) estimating consumption, or quantities of drugs used, and (2) collecting biomarkers, such as hair and urine, from NSDUH respondents.

At the request of ONDCP, the National Research Council (NRC) had convened a committee in 1998 to assess existing data sources and research on drug use and recommend additional data and research to evaluate drug control policies. The committee recommended that methods be developed for acquiring consumption data.[14] Recognizing that this data gap still existed in 2006, ONDCP approached SAMHSA requesting we add consumption questions to NSDUH. The NSDUH team collaborated with ONDCP to convene an expert consultant meeting on June 19, 2007.[15] After the meeting, the NSDUH team worked with a subgroup of the consultants to develop a new module to measure marijuana consumption, but after extensive testing and refinement, it became clear that the module would not produce accurate estimates. The report sent to ONDCP said that more development and testing would be needed before valid and reliable consumption questions could be added to the NSDUH. ONDCP seemed to accept the conclusion and did not push SAMHSA to continue pursuing this research.

Reflecting his longstanding mistrust of NSDUH and other self-report data on drug use, ONDCP Director John Walters said in a May 2006 letter[16] to HHS Secretary Michael Leavitt, "Of particular importance is the collection of biometric data to validate self-reports, without such data policy officials do not have a true understanding of the scope of the problem. To better inform policy decisions, HHS should develop biometric data collection capabilities." In conjunction with the work on consumption, the NSDUH team collaborated with ONDCP to convene an expert consultant meeting on collecting biometric data. This meeting occurred the day after the consumption meeting.[17] The general conclusion of the experts' comments was that there were still unresolved concerns about the accuracy of biometric tests, and SAMHSA should consider conducting periodic validity studies, taking advantage of advances in testing technology as they occur. They recommended against full scale testing of NSDUH respondents.

Several new SAMHSA priorities emerged during the redesign content assessment that began in 2008. Influenced by NSDUH data that showed most people who needed substance abuse treatment did not receive treatment or even feel they needed it, SAMHSA began promoting a practice called Screening, Brief Intervention, and Referral to

Treatment (SBIRT). To address this initiative, the NSDUH team developed new questions asking respondents whether a doctor or other health professional had discussed the respondent's substance use with them in the past twelve months. The ongoing wars in the Middle East raised concerns about the health of returning veterans, so we developed questions on veteran status, service in combat zones, and reserve status. With the proliferation of states legalizing marijuana for medical use, we constructed questions asking marijuana users if any or all of their use in the past twelve months had been recommended by a doctor. Because of SAMHSA's urgent need for data on these topics, these items were added to the questionnaire beginning in 2013.

During the Obama administration, under the direction of Administrator Pamela Hyde, SAMHSA set priorities based on a set of "strategic initiatives." Internal workgroups were created to discuss steps SAMHSA could take to address the initiatives, including collecting new data. The workgroups were interested in adding new NSDUH modules on military families, trauma, and recovery. A question asking if any immediate family members were active duty military was added to the survey in 2015, but modules on trauma and recovery did not emerge so easily. It was difficult to get a consensus on defining these concepts, and even more difficult to translate these definitions into survey questions that respondents could reliably answer. Another barrier was the lack of coordination between the workgroups and the NSDUH team. As of 2018, no trauma module or items had been developed for NSDUH (See discussion of NRC committee in the next section). A set of recovery questions was inserted in the 2018 questionnaire, to provide limited data on this high priority population. The questions simply ask respondents, for substance use and mental health separately, if they think they ever had a problem, and if so, whether they considered themselves to be in recovery. The questions did not undergo cognitive laboratory or field testing.

The marijuana market module, which asks respondents about their purchasing behavior and price paid for marijuana, had been in the survey before 2015, but was dropped as part of the 2015 redesign plan. Although ONDCP had approved the redesign plan, they later indicated that they wanted the module put back in the survey. ONDCP needed these data for periodic studies they conducted on prices and consumption. SAMSHA returned the module to the questionnaire as of 2018. This module, as well as the new recovery questions, was placed at the very end of the ACASI portion of the interview, to minimize the chance that trends for other measures would be affected.

Mental Health Surveillance

The most significant area of interest for expansion of NSDUH, continuing throughout SAMHSA's existence, was mental health, since SAMHSA had never done a parallel mental health survey. Questions on mental health treatment were added in 2000, and a commonly used scale measuring general psychological distress (K-6) was added to the survey in 2001. New modules on depression for youth and adults were added in the 2004 NSDUH. In late 2006, SAMHSA's Center for Mental Health Services (CMHS) expressed interest in collaborating with OAS, to add funds to the NSDUH to obtain state level estimates of the prevalence of serious mental illness (SMI) among adults. These estimates, as well as estimates of serious emotional disturbance (SED) among children, were a requirement of the ADAMHA Reorganization Act, and had been compiled using a flawed estimation model. CMHS was under pressure from Congress to show that they were effectively using their funds to comply with the law. OAS and CMHS convened a group of expert consultants to solicit recommendations on the best approaches for SAMHSA to use for estimating SMI and SED in the future.[18]

The consultants recommended a model-based estimation approach for SMI, using NSDUH. A direct estimation approach, in which each respondent is assessed for SMI, was not considered for NSDUH due to the amount of interview time that would be required. With the model-based approach, a small set of short scales that were highly correlated with SMI, such as K-6, would be included in the questionnaire. A subsample of NSDUH respondents would be recontacted for a longer psychiatric clinical interview to determine their SMI status. Analysts would construct a statistical model based on the clinical interview data to predict whether respondents had SMI, based on responses to the short scales. The procedure had been piloted in NSDUH several years earlier, using just the K-6 data, but it was discontinued due to inadequacy of the K-6 alone to accurately predict SMI. That predictive model had been developed based on a small convenience sample of clinical interviews in one community. With sufficient funding support from CMHS in 2007, more rigorous assessment of prediction models could be done with a larger, nationally representative clinical sample.

After the OAS/CMHS expert meeting, we modified the NSDUH contract to cover the new SMI activity, added the CMHS funds to the contract, and began the developmental work. The study was named the Mental Health Surveillance Study (MHSS). A successful pilot test was conducted in 2007, and the new scales were added in January 2008.

New questions for adults on suicide thoughts, plans and attempts were also added. At the end of each completed adult NSDUH interview, a random selection determined whether or not an adult NSDUH respondent was selected for the follow-up clinical interview. If they were, the interviewer would explain the study and request participation, which entailed a telephone interview by a trained clinician, to be completed within four weeks of the NSDUH interview. If the respondent agreed to the telephone interview, they were given an additional thirty dollars. Approximately 5,500 clinical interviews were completed during 2008–12.[19] SAMHSA began producing annual estimates of the prevalence of SMI after the 2008 survey using a model derived from the 2008 dataset. The model was updated using the full MHSS sample after the 2012 survey.[20] All previously published estimates were revised based on the updated model, and SAMHSA continued to produce SMI estimates in subsequent years.

The interest in expanding NSDUH mental health content continued after the MHSS clinical interviewing ended in 2012. In 2014, SAMHSA and the HHS Office of the Assistant Secretary for Planning and Evaluation (ASPE) jointly funded the NRC formation of a Behavioral Health Measures Standing Committee to study and make recommendations for expanding SAMHSA's behavioral health data collection, focusing on measuring (1) serious emotional disturbance in children, (2) trauma, (3) recovery from a substance use or mental disorder, and (4) specific mental illness diagnoses with functional impairment. The committee held several workshops and prepared reports. Ultimately, the committee did not strongly endorse an expansion of NSDUH to collect these data.[21]

It remains to be seen how far ONDCP will allow its principal source of drug data to shift towards a mental health focus. President Obama's ONDCP Directors Kerlikowske and Botticelli emphasized the public health aspects of the drug problem more than their predecessors. They publicly eschewed the term "War on Drugs," instead describing the drug problem as a public health issue requiring policies driven by data and research. Botticelli himself was the first ONDCP Director in long term recovery from a substance use disorder. After his confirmation in February 2015, he said:

> I am honored by the confidence placed in me by the President and Congress to direct this nation's drug policy. There are millions of Americans – including myself – who are in successful long-term recovery. Our stories can fundamentally change the way our Nation views people with a substance use disorder, which is a disease needing medical treatment like any other disease. As Director, I will continue to advance a science-based drug policy to reduce drug use and its consequences through a balanced approach to public health and public safety.

Notes

1 Because many households do not include a 12–17- or 18–25-year-old resident, approximately 130,000 household screenings were completed each year to obtain the required 45,000 interviews in those age groups.
2 Kroutil, Vorburger, Aldworth, and Colliver, "Estimated Drug Use," 74–87.
3 Obama appointed Gil Kerlikowse as ONDCP Director, Kathleen Sebelius as HHS Secretary, and Pamela Hyde as SAMHSA Administrator. Five months after the expert consultant panel meeting, Obama nominated one of our panel members, Robert Groves, to be Director of the Census Bureau.
4 *USA Today*, December 20, 1990.
5 Mike Jones became project officer in September 2009, and Peter Tice became alternate project officer in March 2011. When Jones left SAMHSA in June 2012, Tice became project officer.
6 SAMHSA attendees included OAS Director Pete Delany and the OAS NSDUH staff.
7 CBHSQ, *NSDUH: 2012 Questionnaire Field Test.*
8 CBHSQ, *NSDUH: 2013 Dress Rehearsal.*
9 CBHSQ, *NSDUH: 2014 and 2015 Redesign.*
10 Early 2015 results revealed high levels of missing data and misreporting among 12–17-year-old respondents, so SAMHSA decided to move marital status back to CAPI in 2016.
11 CBHSQ, *2014 NSDUH: Methodological Resource Book*, Section 15, *Sample Redesign Impact Assessment.*
12 CBHSQ, *2015 NSDUH: Summary of the Effects.*
13 CBHSQ, *2015 NSDUH: Methodological Resource Book*, Section 15, *2015 Questionnaire Redesign Impact Assessment.*
14 NRC, *Informing America's Policy*, 275–6.
15 Experts attending were: Jonathan Caulkins, Deborah Dawson, Dale Hitchcock, Graham Kalton, Patrick O'Malley, Rosalie Pacula, John Pepper, Carol Petrie, Peter Reuter, Bill Rhodes.
16 In accordance with legislation (P.L. 105–277), ONDCP annually provided all agencies involved in drug control with guidance for preparing their annual budgets.
17 Experts attending were: Donna Bush, Yale Caplan, Wayne Ensign, Michael Fendrich, Ron Flegel, Robert Groves, Lana Harrison, Dale Hitchcock, Marilyn Huestis, Clifford Johnson, Timothy Johnson, Graham Kalton, Steven Martin, Patrick O'Malley, Robert Stephenson.
18 Consultants attending the December 2006 meeting included Stephen Buka, Rosanna Coffey, Jane Costello, Robert Friedman, Howard Goldman, Kimberly Hoagwood, Charles Holzer, Ron Honberg, Ron Kessler, Chris Koyanagi, Phil Leaf, Ted Lutterman, Tami Mark, Joe Morissey, Elizabeth Prewitt, Davis Shern, and Alan Zaslavsky. Representatives from SAMHSA, NIMH, NCHS, CDC, ASPE, and Congressional staff also attended.
19 The sample size was achieved with funding support from the National Institute of Mental Health.

20 The predictive model for SMI was also used to generate estimates of "any mental illness" (AMI). SMI reflects the subset of persons with AMI that experience severe impairment as a result of their mental illness. See CBHSQ, "Estimating Mental Illness."

21 Reports, commissioned papers, and meeting information at: http://sites.nationalacademies.org/DBASSE/CNSTAT/Behavioral_Health_Measures_Committee/index.htm.

Wrap-Up for Chapters 11 and 12

Since its inception in 1971, the survey has experienced two major periods of change. The expansion from a small periodic study to a major federal statistical program occurred between 1991 and 1994. A further expansion that increased the sample threefold to produce both national and state estimates occurred during 1999–2002. This expansion occurred at the same time the survey introduced computer-assisted data collection and other upgrades. As was the case for the initial survey, these changes to the survey were largely in response to societal factors: concerns about the effects of rapidly increasing drug use, and the passage of drug legalization laws in states. Other external influences have led to less severe survey changes, such as adding questions to address emerging drug use behaviors or public health problems. But the survey methodology and sample have been relatively stable since 2002, with no major external impetus for a redesign. Overall, the majority of changes were generated internally, by staff working on the survey or by agency leaders. SAMHSA administrators and center directors influenced the questionnaire content, intending to ensure that the survey would produce the data useful to agency's programmatic needs. Questionnaire modules covering mental health, prevention, workplace issues, and treatment are a result of these influences. The survey staff and contractor have consistently had a significant impact on the design of the survey. Numerous innovations were initiated, developed and implemented by them throughout the survey's history, including:

- The sample design was modified to oversample blacks and Hispanics from 1985 through 1998.
- The target population was expanded in 1991 to cover the civilian noninstitutionalized population in all fifty states and the District of Columbia.
- A quarterly sampling plan and continuous data collection began in 1992.

- In 1994, based on a series of methodological studies, the survey fielded an improved questionnaire and estimation procedures.
- The paper questionnaire was converted to CAI in 1999.
- The state-based sampling plan was developed, in coordination with the small area estimation methodology to produce the state estimates beginning in 1999.
- Incentive payments were introduced in 2002.
- Survey respondents' confidentiality protections were strengthened in 2004 with CIPSEA and further solidified in 2006 with the designation of OAS as a federal statistical unit.
- A revised sample design was employed in 2014 to update age and state allocations and clustering to improve the precision of many key national estimates and improve data collection efficiency.
- In conjunction with the 2014 sample design update, a revised questionnaire and other data collection materials to improve data quality and relevance were implemented in 2015.

It's notable that the last two items were the result of a nearly decade-long process of outreach, development, and testing of new methods and questions that was initiated and managed by the OAS NSDUH team. There was no external push for a redesign, although communication with and approvals by the relevant administration officials was frequent. This transparency (including explicit ONDCP approval of the plan and the early release of the preliminary trend break report) undoubtedly contributed to the unhindered release of the 2015 data.

The process for accomplishing the 2014–15 redesign was based on the belief that ongoing surveys need to be continually evaluated and periodically updated if appropriate. This idea applies even to surveys like NDSUH that are used to track trends. However, if trend measurement is a priority, the interval between major redesigns may need to be long, with very limited changes between redesigns. Also, if possible, trend measurement may be preserved with some kind of bridge between old and new data as part of the redesign implementation. But as we have seen on multiple occasions with NSDUH, sometimes seemingly minor changes to the questionnaire or other aspects of the survey can change the way respondents answer questions, causing trend breaks.

Most of the changes made by the NSDUH survey team were intended to improve the quality of the data and the potential for producing valuable policy and epidemiological research studies. The changes helped to maintain the statistical integrity of the survey. But surveys can sometimes encounter difficult policies or hostile leadership that endanger statistical integrity. NSDUH has encountered this from the highest levels of government, such as when President Nixon told Raymond Shafer that he wanted the commission report to come out strongly against marijuana, or when ONDCP demanded immediate sample size increases and major questionnaire changes. But attacks on a survey's statistical integrity can come from within the agency or even within the office responsible for the survey. OAS recently experienced two directors, neither of which had backgrounds in survey research, whose policies and management diminished NSDUH's integrity. During their tenures we experienced cuts in data analysis, less transparency, lack of support for statistically sound practices, exclusion of statisticians from discussions of data issues, discontinuation of the permanent expert consultant panel, and nearly a disastrous major reduction of the questionnaire. There were also staff reductions, low morale, and loss of productivity. The project recovered from that era and has maintained statistical integrity, but there are concerns in CBHSQ and throughout the federal statistical system about the anti-science policies of the Trump administration.[1] When an administration issues rules prohibiting its major public health agency from using terms like "science-based," and regularly cites erroneous "facts," one has to wonder about the impact that administration will have on the design and reporting of results from the surveys they conduct.[2]

Notes

1 Pierson, "ASA, with Statistical Community."
2 Sun and Eilperin, "CDC Gets a List of Forbidden Words, Including 'Diversity' and 'Transgender,'" *Washington Post*, December 16, 2017, A4.

13 Lessons Learned and Future Challenges

The story of the National Survey on Drug Use and Health is one of growth, with a mix of success and conflict. An epidemic of illicit drug use emerged in the United States during the 1960s. Despite the lack of data available to characterize the nature and scope of the problem as it emerged, there was a consensus that government action was needed to halt the increasing use. A national survey to track illicit drug use was authorized. Drug use continued to increase during the 1970s and persisted thereafter. As government leaders began using the survey's data, they recognized its limitations and invested in enhancements. The survey evolved into a valuable tool for policymaking and research, and an important component of the nation's federal statistical system.[1] The story of this evolution contains useful lessons for survey statisticians, government leaders, policymakers, and other users of statistical data.

Improving Communication While Maintaining Statistical Integrity

The survey's history is littered with examples of communication breakdowns that led to delays, wasted effort, and bad decisions. Sometimes these communication difficulties were due to bureaucratic barriers or the personalities and management styles of the people involved. Often the underlying cause was the fundamentally different goals and points of view between survey statisticians and agency leaders. Agency leaders want numbers they can use to help them make decisions, develop policies, or to bring attention to a problem. They may want data to justify decisions they've made or policies they espouse, or to support their claims of success. Often they need information quickly, in an easily digestible form, uncluttered by extra information about caveats or on the accuracy of the estimates. If the data they need are not available, they want it produced right away. Statisticians, on the other hand, strive for careful assessment and reporting of data limitations, and information on the precision of estimates. They use this supplementary information to help interpret the data objectively and assess appropriateness for uses. Sometimes this careful data production takes time. Some estimates may be deemed too unreliable to use. If the data are not yet collected, it may take several years to develop new questions, collect the data, and produce estimates.

So how can agency leaders and survey statisticians work together to ensure that statistical data from surveys and other sources are effectively used to help solve the nation's substance abuse and other problems? My NSDUH experience points to at least three areas for improvement: education, empathy, and respect. Agency leaders would benefit from some courses or short training modules on statistics and survey methodology. Managers of statistical units and data programs should be required to have training and experience in relevant areas of statistics, such as survey research. In situations where they don't have this expertise, these managers need to make a concerted effort to understand the motivations and principles that statisticians bring to their work, and appreciate their unique contribution and expertise. Agency leaders should ensure that statisticians are an integral part of planning and decision-making on data issues. Agency heads as well as survey managers and staff should rely on the guidelines developed by the Office of Management and Budget[2] and the National Research Council.[3]

There are also ways that statisticians could help improve communication with agency leaders. Statisticians should improve their communication skills (both written and verbal), learning how to better explain statistical issues and principles in simple, concise ways that non-statisticians can understand. Communication proficiency should be emphasized in statistics degree programs. With this knowledge, statisticians will be equipped to explain how the adherence to core statistical principles and standards will be beneficial to achieving leaders' goals. Statisticians should also become familiar with policy and research issues and societal shifts associated with the data they produce and analyze, so they can better understand the motivations and concerns of agency heads and other political leaders, and respect the challenges they face in making policy decisions in the presence of many competing priorities. With a thorough understanding of why the data are needed and how the data will be used, statisticians involved in all aspects of a study can effectively contribute towards achieving the goals of the study.

The Value of Methodological Research

Methodological research has been an important part of NSDUH throughout its history.[4] Studies have evaluated the methods used in the survey and discovered better ways to collect and analyze data.[5] Experiments comparing the effects of data collection methods on the reporting of substance use, and in-depth comparisons with data from other sources have helped analysts better understand how different methodologies can give different estimates of youth substance use,[6] substance use disorders,[7] cigarette use,[8] mental health,[9] and treatment.[10] Armed with this knowledge, analysts can prepare insightful reporting of trends and patterns in substance abuse, interpreting NSDUH results in the context of findings from other surveys collecting similar data. Although the primary purpose of methodological studies was to improve the collection and analysis of NSDUH data, the findings have also been useful to managers and statisticians working on other surveys. Detailed reports on response rate patterns, data collection and estimation methods, tests of new methods and their effects on data, sampling errors associated with prevalence estimates, and other information about the survey methodology have been made available every year when data are released.[11] Maintaining an ongoing methodological evaluation and research program and making these results widely available helped build confidence in the survey itself. The program demonstrated rigor, openness to scrutiny, and statistical integrity.

Underpinning the concept of methodological research is the acknowledgement of the science of survey research. It is a recognition that all aspects of a survey design can influence the resulting data, and these aspects can be experimentally manipulated to measure their effects. These design features are all interrelated and include a variety of influences such as the sample selection, the wording of lead letters, the training of interviewers, the structure and wording of questions, the coding and processing of raw data, and the calculation of estimates and their precision. Analysts can measure the individual effects of these features through methodological research using experimental designs to determine the best methods to apply in a particular survey. An ongoing program of methodological research is a critical component of a continuing survey. It can monitor the performance of survey procedures and catch emerging problems, as well as compile findings to prepare for a future redesign.

Managing Survey Redesigns

Communication and methodological research are also critical factors in developing a survey design and in modifying the survey design to stay relevant and modern. Data users and agency leaders should be consulted early and regularly in the design development phase, to make sure the survey sample and questionnaire content reflect the goals and data needs of its ultimate consumers. In an ongoing survey, often there is a choice between leaving the design intact to maintain comparability versus changing the design to improve the quality and utility of data. The pros and cons of these alternatives should be explained to decision-makers. This may involve communicating complex and sometimes tenuous results from methodological research. For example, when the NSDUH team was asked to add new questions to the survey, we sometimes had to explain the potential for context effects, whereby a new question inserted at one point in the interview could impact how respondents answer questions that follow that insertion. We cited results from methodological studies showing how this can occur.[12]

It is also important that the methods teams involved in research and redesign maintain good communication with the staff responsible for the ongoing survey operations, including data collection, processing, and reporting. This is true whether the methods and the operations staffs are integrated or are separate units, but is more challenging in the latter case.

Effective Use of Consultants

Throughout its history, the NSDUH has benefitted from the advice of expert consultants. We contacted these experts either individually to ask for help on specific issues, or as part of a group that met to discuss issues with the NSDUH team. Meetings with invited consultants were convened numerous times to help guide decision-making on all aspects of the project. A great deal of careful planning and effort was needed to make these meetings productive. Obviously, choosing well-respected, thoughtful consultants was an important first step. Often we called upon certain experts multiple times, including a permanent expert panel that met with us regularly from 1999 to 2008. These experts became familiar with the details of the survey and the NSDUH staff, providing continuity in the advice they contributed. Regardless of the particular areas of expertise of the participating consultants, preparing them for the meeting was critical. Concise, informative reading materials explaining the goals of the meeting and background descriptions of the topics to be discussed were sent to consultants prior to the meeting. The goals and background were usually reinforced with summary presentations by NSDUH staff at the start of the meeting. A skilled chairperson was needed to keep the discussions focused on the goals, elicit responses from consultants, and make sure everyone had opportunities to speak. Finally, careful note-taking and preparation of a meeting summary was a priority, providing a document that could be referred to later in describing what influenced the agency to make decisions on the survey design or other aspects of the project.

Making Connections across the Survey Research Community

A great resource for statisticians involved with surveys is their counterparts working on other surveys. It's rare to encounter a problem or a challenge in your own work that has not been faced by staff in another survey. Countless times in approaching issues on NSDUH, someone on the team would ask what they do in the Health Interview Survey, the Monitoring the Future study, the National Comorbidity Study, or the National Survey of Family Growth. We learned from those surveys, by reading their reports and talking with their leaders and staffs. We got to know them, and became collaborators with some of them. I found the survey research community, federal and nonfederal, to be a very cooperative, approachable, helpful group. Most statisticians are eager to share their expertise and experiences with colleagues from other studies who confront them with difficult problems, or just want to learn about their methods. Many in the survey research community have valuable knowledge that has not been published, so it is only available through direct contact.

One way to learn about other surveys, build networks, and to keep informed about the latest advances in the science of survey research in general is to participate in professional conferences such as the annual Joint Statistical Meetings and the American Association of Public Opinion Research Conference. The benefits to the NSDUH and its staff (NIDA, SAMSHA, and contractors) of attending and presenting at conferences were clear to me. At these conferences, statisticians working on other surveys attended presentations by NSDUH staff and became interested in NSDUH, and sometimes offered advice based on their experiences. We met colleagues who had done similar research or faced the same problems and learned how they handled them. Making these connections also helped build a pool of experts we could later draw from to form expert consultant panels. By presenting methodological information on the NHSDA and answering questions about the survey in a public forum that included some of the top statisticians in the world, the project team built a reputation of openness, rigor, and competence. The survey gained respect, as did the organizations responsible for the survey. The organizations also benefitted by having their staff gain knowledge from attending various sessions at the conference. Staff learned about the latest findings and methodologies in survey research, and connected with statisticians who could later be called upon for advice.

Another way for statisticians to develop long-standing connections with experienced survey experts from whom they can learn is to spend some time working in a large statistical organization (government or private) that has a cadre of experts and runs a variety of types of surveys. This is of course best done in the early stages of a career. Before I went to work at NIDA and was given the responsibility to manage the NSDUH, I spent four years working at the National Center for Health Statistics. At NCHS, I was assigned to work on the design and analysis of several of their major national surveys, including the National Health Interview Survey. During those four years I had opportunities to work with and learn from many excellent NCHS statisticians. I observed how procedures taught in statistics courses, such as selecting a representative sample and creating valid estimates from it, or testing whether the difference between two estimates is statistically significant, are actually carried out in the real world. I participated in developing questionnaires, training interviewers, observing real interviews, developing editing specifications and weights, and writing reports of findings. I learned the importance of documenting survey design characteristics, along with all of their caveats, and accounting for them when reporting the results, and making sure the results are reported accurately. That means not only the right numbers, but also the appropriate interpretation of the numbers. My exposure to and involvement in the day-to-day operations of a large statistical agency gave me a good background for managing the NSDUH.

The Real Value of Surveys: Findings

The most important purpose of any survey is the findings it provides. Therefore, every aspect of survey work should occur with analysis plans in mind. Statisticians developing a survey design must first have a clear understanding of the goals of the survey, including what kinds of policy and research questions the survey data is supposed to answer, and what kinds of estimates and analyses are expected. Of course the content of the questionnaire needs to reflect the analysis plan, but so should the sample design, method of data collection, editing, imputation, weighting, and data file construction.

Data collection and processing can be expensive. But all of it is worthless unless the data are used. Reports of the main findings, data tables providing detailed results, and studies focused on specific populations or issues should be produced. I found that producing a comprehensive report of the main findings and methodological details first, and having it available when the initial survey results were released, tended to lead to accurate and unbiased reporting of the findings by government leaders and the press.[13]

The value of a survey can be increased by the findings generated by independent researchers. Often these researchers have unique ideas for analysis of the survey data that can produce important results, and they obtain financial support through grants to carry out this secondary analysis. But this is only possible if the survey organization makes the data available in a form conducive to this type of study. Accomplishing that goal was a priority of the NSDUH team. The potential benefit (increasing the value of the survey) far outweighed the cost of preparing the files and setting up access systems. But legal and ethical concerns had to be considered. Protecting the confidentiality of NSDUH respondents required disclosure treatments to be applied to public use data files. Provision of confidential data files to outside researchers could only be done using specific arrangements allowed by law, and in a way that gave equitable access to all researchers.[14]

Next for NSDUH

Future NSDUH survey design and analysis teams will undoubtedly have to deal with new issues that could impact the data and indicate the need for survey design changes. Marijuana laws are rapidly changing, so new items may be needed to track changes in use by legal and medical status. Perhaps marijuana use in states that have legalized use should not be lumped with "illicit drug use" in national estimates. Methodological studies may be needed to determine whether NSDUH respondents who use marijuana are more likely to admit their use if they live in an area where use is legal. If this were to occur, it would complicate comparisons of rates of marijuana use over time and across states. Questions could be added to track use of new nicotine products and other modes of substance ingestion such as vaping and hookahs. Pressure to expand mental health data collection in NSDUH will probably continue, particularly given the dearth of national data on children. Efforts to measure recovery and trauma are likely to continue. With all of the focus on expanding what NSDUH collects, and little support for cutting any topic areas from the questionnaire, SAMHSA may soon have to consider mounting a separate survey to capture some of the new data. Perhaps too much is being asked of one survey, and it would be better to limit NSDUH's scope to be closer to its original purpose, which it has historically and uniquely done best: describe the nature and extent of illicit drug use.

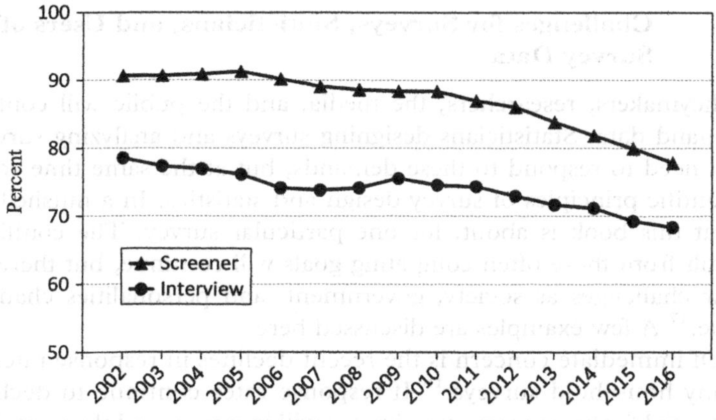

Figure 13.1 NSDUH response rates, by year.

New ways of collecting data should be explored to adapt the survey to societal changes that may be affecting survey response rates. NSDUH, like many other federal surveys, has recently experienced alarming declines in response rates (see further discussion in the next section). Figure 13.1 shows that the proportion of households that completed the short household screening questionnaire each year had been steady at more than 90 percent through 2006, but then began to decline gradually, reaching 86 percent in 2012. By 2016, the household screening response rate had dipped to 78 percent. The proportion of sample persons completing the interview followed a similar pattern, declining from 73 percent in 2012 to 68 percent in 2016. Could these declines in response rates be causing a distorted picture of trends in substance use? More methodological research is needed to address the response rate.

Not long after the 2015 partial redesign was in place, the NSDUH team, headed by Jonaki Bose,[15] began discussing the next redesign. The NSDUH contract to conduct the 2018–22 surveys included support for the redesign activity. It was awarded to RTI in December 2016. CBHSQ and RTI began discussing plans for relevant methodological development in 2017, and formal outreach for suggestions and priorities for the next redesign began with a Federal Register notice in April.[16] Some of the obvious areas for exploration and potential modification include updating the substance use disorder module from DSM-4 to DSM-5 criteria, adding questions on emerging modes of administration of tobacco, marijuana, and other substances, and increasing the interview incentive amount from thirty dollars to account for inflation and to improve response rates.

Challenges for Surveys, Statisticians, and Users of Survey Data

Policymakers, researchers, the media, and the public will continue to demand data. Statisticians designing surveys and analyzing survey data will need to respond to these demands, but at the same time adhere to scientific principles of survey design and statistics. In a nutshell, that is what this book is about, for one particular survey. The conflicts that result from these often competing goals will continue, but there will be new challenges as society, government, and personalities change over time.[17] A few examples are discussed here.

Of immediate concern is the recent declines in response rates across many household surveys.[18] If response rates continue to decline, the potential for bias in survey estimates will increase, and the confidence in survey data will erode. Statistical integrity will be diminished. While the reasons for the declines are not fully understood, some factors that may play a role are growing concerns about privacy, intrusions on people's time, preponderance of marketing and political pitches (sometimes posing as surveys) that cause people to tune them out, and the availability of barriers such as call screening and gated communities. Approaches that statisticians and survey managers might consider to improve response rates, or at least slow the decline, include reducing the length of interviews (minimizing respondent burden), using procedures that make interviews a more positive experience for respondents, and giving respondents incentive payments. As social media outlets evolve, they may also become a mechanism for survey data collection.[19] Another possibility is using multiple modes, to allow respondents to participate using their preferred method, such as by mail, phone, in-person interview, or via the Internet.[20] The multi-mode approach could be problematic for surveys like NSDUH that collect data known to be strongly affected by the choice of the data collection mode. To combat the bias caused by declining response rates, statisticians will need to develop improved statistical adjustments, possibly using auxiliary data that can be incorporated into weighting algorithms. This auxiliary data could come from new items answered by household screening respondents, from interviewer observations about the neighborhood, family, or interview process (paradata), or from external data sources linked to the survey data.[21] More research is needed to better understand the causes of nonresponse and to determine effective strategies for improving response rates. The research, and some of the proposed solutions, will add costs to survey projects, so statisticians will need to prepare justifications based on the benefits in terms of improved data quality.

Analysts should look for new opportunities and approaches for survey research studies. Linking survey data with external data sources has great potential for other purposes besides nonresponse adjustments. These linkages can be made either at an aggregate or individual level.[22] For example, research studies have linked NSDUH data on state policies and local area characteristics to explore the impact of those policies and characteristics on health care and behaviors.[23] Survey respondent data have also been linked with death records to study the relationship between health behaviors and mortality.[24] Although there are tricky confidentiality and data access issues to overcome in such studies, it seems that these kinds of linkages will become more common as new electronic data files become available.[25] They are a cost-effective way to enhance the content of a survey's data, and the variety of available data sets that could be linked, including administrative data sets,[26] seems to be increasing.

Finally, the increasing use of social media, while providing potential new ways to collect data, also has implications for communicating survey results. There has been a shift in how the public obtains news and other information, away from printed newspapers and network television news broadcasts, towards social media.[27] Methods used by survey organizations to disseminate survey results have also changed, with less reliance on printed reports, and more use of web-only documents and social media vehicles such as blogs, Facebook, and Twitter. But the wide availability and usage of these outlets also makes it easy and efficient for any organization or individual to disseminate information. People may quote from their own legitimate published studies or from other valid studies. Or they may parrot the results from poorly designed studies, or give numbers that were simply fabricated. Politicians or consumer products marketers like to cite data to support their claims. Numbers seem to make them more believable. Given their ability to send messages and postings to potentially millions of receptive readers, false or misleading research findings and data can go viral. More than ever, the public will need to be wary of survey and other research findings they encounter through social media. And statisticians should play a role in guiding the public through the mix of valid data and fake research news, especially during an era in which government leaders can actually be the producers and supporters of fake research news and critics of objective, science-based findings. In closing, I hope statisticians will continue to be warriors in the battles for statistical integrity in their work, and that the public will increasingly recognize and appreciate statistical integrity when reading about survey results.

Notes

1 OMB, *Statistical Programs.*
2 OMB, *Statistical Policy Directive Number 2*; OMB, *Statistical Policy Directive No. 4.*
3 NRC, *Principles and Practices.*
4 CBHSQ, *NSDUH: Summary of Methodological Studies.*
5 Turner, Lessler and Gfroerer, *Survey Measurement*; Gfroerer, Eyerman, and Chromy, *Redesigning an Ongoing.*
6 Gfroerer, Wright, and Kopstein, "Prevalence of Youth," 19–30; SAMHSA, *Comparing and Evaluating.*
7 Grucza et al., "Discrepancies in Estimates," 623–9.
8 Ryan, Trosclair, and Gfroerer, "Adult Current Smoking."
9 Hedden et al., *Comparison of NSDUH.*
10 Batts et al., *Comparing and Evaluating.*
11 CBHSQ, *NSDUH: Methodological Resource Book.*
12 Colpe et al., "Screening for Serious Mental," 210–11.
13 For example: SAMHSA, *Results from the 2013.*
14 SAMHSA, *The NSDUH Report: Accessing.*
15 Subsequent to my retirement in January 2014, Art Hughes became acting chief of the Populations Survey Branch. Jonaki Bose was appointed as permanent branch chief in August 2015.
16 Federal Register Vol. 82, No. 79.
17 National Science Foundation, *The Future of Survey Research.*
18 National Research Council, *Nonresponse in Social Science.*
19 Hill, Dean, and Murphy, *Social Media, Sociality.*
20 Groves et al., *Survey Methodology*, 153–5, 163–5.
21 Biemer and Peytchev, "Using Geocoded Census Data," 24–44.
22 Johnson, *Handbook of Health.*
23 Farrelly et al., "Comprehensive Examination," 549–55; Compton et al., "Unemployment and Substance," 350–53.
24 Muhuri and Gfroerer, "Mortality Associated with Illegal," 155–64.
25 National Academies, *Innovations in Federal Statistics.*
26 OMB, "Guidance for Providing."
27 Pew Research Center, "News Use across Social," 9.

Appendix

Table 1 *Drug surveys with national samples, using paper and pencil interviewing, 1971–1998*

Year	Data Collection Period	Contractor	Sample Size	Interview Response Rate (%)	Notes on Design
1971	September–October	RAC	3,466	81	
1972	September–October	RAC	3,291	76	
1974	November 74–March 75	GWU-RAC	4,023	83	
1976	January–May	GWU-RAC	3,576	84	
1977	March-July	GWU-RAC	4,594	85	
1979	August 79–January 80	GWU-RAC	7,224	86	Oversample of rural areas
1982	March–July	GWU-RAC	5,624	84	
1985	June–December	ISR-UK	8,038	83	Start of oversampling of blacks, Hispanics (1985–1998)
1988	September 88–February 89	RTI	8,814	74	
1990	March–June	RTI	9,259	82	Oversample of DC metro area
1991	January–June	RTI	32,594	84	Oversample of six metro areas; target population redefined as civilian noninstitutional, including AK and HI
1992	January–December	RTI	28,832	83	Oversample of six metro areas; quarterly (continuous) sampling begins

Table 1 (*cont.*)

Year	Data Collection Period	Contractor	Sample Size	Interview Response Rate (%)	Notes on Design
1993	January–December	RTI	26,489	79	Oversample of six metro areas
1994	January–December	RTI	17,809	78	New questionnaire and editing rules begin; separate sample of 4,372 given "old" questionnaire for creating adjustment factors; rural oversample (1994 only)
1995	January–December	RTI	17,747	81	
1996	January–December	RTI	18,269	79	
1997	January–December	RTI	24,505	78	Oversample of AZ and CA
1998	January–December	RTI	25,500	77	Oversample of AZ and CA

Note: Federal sponsors of the survey were the National Commission on Marihuana and Drug Abuse (1971–72), the National Institute on Drug Abuse (1974–92), and the Substance Abuse and Mental Health Services Administration (1993 and after). Contractors were Response Analysis Corporation (RAC); George Washington University (GWU); Institute for Survey Research, Temple University (ISR); University of Kentucky (UK); and Research Triangle Institute (RTI).

Table 2 *Drug surveys with national and state samples, using computer-assisted interviewing, 1999–2016*

Year	Sample size	Screening Response Rate (%) (weighted)	Interview Response Rate (%) (weighted)	Notes on Design
1999	66,706	90	69	Extra 2,500 youth interviews funded by HHS; separate national sample of 13,809 given 1998 PAPI questionnaire
2000	71,764	93	74	Extra 2,500 youth interviews funded by HHS. Hair/urine collected on subsample of 2,000 12–25-year- olds
2001	68,929	92	73	Hair/urine collected on subsample of 2,000 12–25 year olds; incentive experiment on subsample of 4,000 (January-June)
2002	68,126	91	79	Survey name changed to NSDUH; began giving thirty dollar incentive to all respondents
2003	67,784	91	77	
2004	67,760	91	77	Contact materials cite CIPSEA for the first time
2005	68,308	91	76	
2006	67,491	90	74	
2007	67,377	89	74	
2008	67,928	89	74	MHSS clinical interviews on subsample of 1,500 adults
2009	68,007	88	76	MHSS clinical interviews on subsample of 500 adults
2010	67,804	88	75	MHSS clinical interviews on subsample of 500 adults
2011	70,109	87	74	MHSS clinical interviews on subsample of 1,500 adults; supplemental sample of 2,000 in Gulf Coast region
2012	68,309	86	73	MHSS clinical interviews on subsample of 1,500 adults
2013	67,838	84	72	
2014	67,901	82	71	Shifts in age and state sample allocations
2015	68,073	80	69	Revised contact materials and questionnaire
2016	67,942	78	68	

Note: All data collection was done January through December each year, with quarterly samples. The contractor was RTI, International for all surveys shown.

Note: Weighted interview response rate tends to be lower than the unweighted rate shown in Table 1, because adults over age 26 are weighted more heavily, and they tend to have lower response rates than persons under age 26.

Bibliography

Abelson, Herbert, Cohen, Reuben, Heaton, E., and Suder, C. *Public Attitudes Toward and Experiences With Erotic Materials*. Technical Report of the Commission on Obscenity and Pornography, Volume VI. Washington, DC: US Government Printing Office, 1970. Available online at Hathi Trust Digital Library. http://babel.hathitrust.org/cgi/pt?id=mdp.39076006957836;view=1up;seq=1.

Adams, Edgar H. and Gfroerer, Joseph C. "Elevated Risk of Cocaine Use in Adults." *Psychiatric Annals* 18(9) (September 1988): 523–7.

Adams, Edgar H., Rouse, Beatrice A., and Gfroerer, Joseph C. "Populations at Risk for Cocaine Use and Subsequent Consequences." In *Cocaine in the Brain*, edited by Nora D. Volkow and Alan C. Swann, 25–41. New Brunswick, NJ: Rutgers University Press, 1990.

Anderson, Margo J. *The American Census: A Social History, Second Edition*. New Haven, CT: Yale University Press, 2015.

Batts, Kathryn, Pemberton, Michael, Bose, Jonaki, Weimer, Belinda, Henderson, Leigh, Penne, Michael, Gfroerer, Joseph, Trunzo, Deborah, and Strashny, Alex. *Comparing and Evaluating Substance Use Treatment Utilization Estimates from the National Survey on Drug Use and Health and Other Data Sources*. CBHSQ Data Review. Rockville, MD: Substance Abuse and Mental Health Services Administration, 2014.

Biemer, Paul, and Peytchev, Andy "Using Geocoded Census Data for Nonresponse Bias Correction: An Assessment." *Journal of Survey Statistics and Methodology* 1(1) (2013): 24–44.

Bennett, William. "Penalties Must Be Harsh to Curtail Drug Use." In *Drug Abuse*, edited by Kelly Barth. Greenhaven Press, 2007.

Bray, Robert M., and Marsden, Mary Ellen, eds. *Drug Use in Metropolitan America*. Thousand Oaks, CA: Sage Publications, 1999.

Bregger, John E. "The Current Population Survey: A Historical Perspective and BLS' Role." *Monthly Labor Review*, June 1984: 8–14.

Bush, George. "Remarks at a White House Briefing on Drug Abuse Statistics," December 19, 1990. Online by Gerhard Peters and John T. Woolley, The American Presidency Project. www.presidency.ucsb.edu/ws/?pid=19169.

Card, David. "Origins of the Unemployment Rate: The Lasting Legacy of Measurement without Theory." http://davidcard.berkeley.edu/papers/origins-of-unemployment.pdf

Carnevale, John, and Murphy, Patrick. "Matching Rhetoric to Dollars: Twenty-Five Years of Federal Drug Strategies and Drug Budgets." *Journal of Drug Issues* 29(2) (1999): 299–322.

Center for Behavioral Health Statistics and Quality. *National Survey on Drug Use and Health (NSDUH): Summary of Methodological Studies, 1971–2014*. Rockville, MD: Substance Abuse and Mental Health Services Administration, 2014.

National Survey on Drug Use and Health: 2012 Questionnaire Field Test Final Report. Rockville, MD: Substance Abuse and Mental Health Services Administration, 2014.

National Survey on Drug Use and Health: 2013 Dress Rehearsal Final Report. Rockville, MD: Substance Abuse and Mental Health Services Administration, 2014.

Behavioral Health Trends in the United States: Results from the 2014 National Survey on Drug Use and Health. Rockville, MD: Substance Abuse and Mental Health Services Administration, 2015.

National Survey on Drug Use and Health: 2014 and 2015 Redesign Changes. Rockville, MD: Substance Abuse and Mental Health Services Administration, 2015.

Estimating Mental Illness Among Adults in the United States: Revisions to the 2008 Estimation Procedures. Rockville, MD: Substance Abuse and Mental Health Services Administration, 2015.

2014 National Survey on Drug Use and Health: Methodological Resource Book (Section 15, Sample Redesign Impact Assessment, Final Report). Rockville, MD: Substance Abuse and Mental Health Services Administration, 2016.

2015 National Survey on Drug Use and Health: Summary of the Effects of the 2015 NSDUH Questionnaire Redesign: Implications for Data Users. Rockville, MD: Substance Abuse and Mental Health Services Administration, 2016.

2015 National Survey on Drug Use and Health: Methodological Resource Book (Section 15, 2015 Questionnaire Redesign Impact Assessment, Final Report, Volume 1). Rockville, MD: Substance Abuse and Mental Health Services Administration, 2017.

Chavez, Nelba. Report to Congress, Conference Appropriations Committee. "Expansion of the National Household Survey on Drug Abuse and Plans to Improve Substance Abuse Services." Rockville, MD: Substance Abuse and Mental Health Services Administration, February 1998.

Clayton, Richard R., and Voss, Harwin L. *A Composite Index of Illicit Drug Use*. Rockville, MD: National Institute on Drug Abuse, 1981.

Colpe, Lisa, Epstein, Joan, Barker, Peggy, and Gfroerer, Joseph "Screening for Serious Mental Illness in the National Survey on Drug Use and Health (NSDUH)." *Annals of Epidemiology* 19(3) (2009): 210–11.

Compton, Wilson, Gfroerer, Joseph, Conway, Kevin, and Finger, Matthew. "Unemployment and Substance Outcomes in the United States 2002–2010." *Drug and Alcohol Dependence* 142 (2014): 350–3.

Compton, Wilson M., and Volkow, Nora D. "Major Increases in Opioid Analgesic Abuse in the United States: Concerns and Strategies." *Drug Alcohol Dependence* 81(2) (2006): 103–7.

Domestic Council Drug Abuse Task Force. *White Paper on Drug Abuse.* September 1975. Washington, DC: US Government Printing Office, 1975.

DuPont, Robert L. "Present at the Creation–NIDA's First Five Years." *Drug and Alcohol Dependence* 107 (2010): 82–7.

Farrelly Matthew, Loomis, Brett, Han, Beth, Gfroerer, Joseph, Kuiper, Nicole, Couzins, G. Lance, Dube, Shanta, and Caraballo, Ralph A. "Comprehensive Examination of the Influence of State Tobacco Control Programs and Policies on Youth Smoking." *American Journal of Public Health* 103 (March 2013): 549–55.

Federal Register Volume 63, No. 120. June 23, 1998. pages 34191–92.

www.federalregister.gov/documents/1998/06/23/98-16616/agency-information-collection-activities-proposed-collection-comment-request

Federal Register Volume 82, No. 79. April 26, 2017. pages 19247–48.

www.federalregister.gov/documents/2017/04/26/2017-08400/request-for-comment-on-the-nsduh-redesign.

Fishburne, Patricia, Abelson, Herbert, and Cisin, Ira *National Survey on Drug Abuse: Main Findings: 1979.* Rockville, MD: National Institute on Drug Abuse, 1980.

Fowler, Floyd J. *Survey Research Methods.* Thousand Oaks, CA: Sage Publications, 2014.

General Accounting Office, *Drug Use Measurement. Strengths, Limitations, and Recommendations for Improvement.* GAO/PEMD-93–18. June 1993

Report to the Committee on Government Operations, House of Representatives. *Drug Control: Reauthorization of the Office of National Drug Control Policy.* GAO/GGD-93–144, September 1993.

Gerstein, Dean, and Harwood, Henrick. *Treating Drug Problems, Volume 1.* Washington, DC: National Academy Press, 1990.

Gfroerer, Joseph. "Influence of Privacy on Self-reported Drug Use by Youths." In *Self-Report Methods of Estimating Drug Use: Meeting Current Challenges to Validity,* edited by Beatrice Rouse, Nicholas Kozel, and Louise Richards, 22–30. NIDA Research Monograph 57. Rockville, MD: National Institute on Drug Abuse, 1985.

"Correlation Between Drug Use by Teenagers and Drug Use by Older Family Members." *American Journal of Drug and Alcohol Abuse* 13(1 & 2) (1987): 95–108.

Gfroerer, Joseph and Brodsky, Marc. "Incidence of Illicit Drug Use in the United States, 1962–1989." *British Journal of Addiction* 87 (1992): 1345–51.

Gfroerer, Joseph, and Epstein, Joan. "Marijuana Initiates and Their Impact on Future Drug Abuse Treatment Need." *Drug and Alcohol Dependence* 54 (1999): 229–37.

Gfroerer, Joseph, Eyerman, Joe, and Chromy, James, eds. *Redesigning an Ongoing National Household Survey: Methodological Issues.* Rockville, MD: SAMHSA, 2002.

Gfroerer, Joseph, Hughes, Arthur, and Bose, Jonaki. "Sampling Strategies for Substance Abuse Research." In *Research Methods in the Study of Substance Abuse,* edited by Jonathan VanGeest, Timothy Johnson, and Sonia Alemagno, 65–80. Springer International Publishing, 2017.

Gfroerer, Joseph, Hughes, Arthur, Chromy, James, Heller, David, and Packer, Lisa "Estimating Trends in Substance Use Based on Reports of Prior Use in a Cross-sectional Survey." In *Eighth Conference on Health Survey Research Method,* edited by Steven B. Cohen and James M. Lepkowski, 29–34. Hyattsville, MD: National Center for Health Statistics, 2004.

Gfroerer, Joe and Kennet, Joel "Collecting Survey Data on Sensitive Topics: Substance Use." In *Handbook of Health Survey Methods,* edited by Timothy Johnson, 447–514. Hoboken, NJ: John Wiley and Sons, 2014.

Gfroerer, Joseph, Lessler, Judith and Parsley, Teresa . "Studies of Nonresponse and Measurement Error in the NHSDA." In *The Validity of Self-Reported Drug Use: Improving the Accuracy of Survey Estimates,* edited by Lana Harrison and Arthur Hughes, 273–295. NIDA Research Monograph 167. Rockville, MD: National Institute on Drug Abuse, 1997.

Gfroerer, Joseph, Wright, Douglas, and Kopstein, Andrea "Prevalence of Youth Substance Use: The Impact of Methodological Differences Between Two National Surveys." *Drug and Alcohol Dependence* 47 (1997): 19–30.

Goldberg, Peter. "The Federal Government's Response to Illicit Drugs, 1969–1978." In *The Facts About Drug Abuse.* The Drug Abuse Council. New York: Free Press, 1980.

Grim, Ryan. *This Is Your Country on Drugs: The Secret History of Getting High in America.* Hoboken, NJ: John Wiley and Sons, 2009.

Groves, Robert M. and Couper, Mick P. *Nonresponse in Household Interview Surveys.* New York, NY: John Wiley & Sons, 1998.

Groves, Robert M., Fowler, Floyd J. Couper, Mick P. Lepkowski, James M. Singer, Eleanor, and Tourangeau, Roger. *Survey Methodology.* New York, NY: John Wiley and Sons, 2004.

Grucza, Richard, Abbacchi, Anna, Przybeck, Thomas, and Gfroerer, Joseph. "Discrepancies in Estimates of Prevalence and Correlates of Substance Use and Disorders Between Two National Surveys." *Addiction* 102 (2007): 623–629.

Han, Beth, Gfroerer, Joseph Colliver, James D., and Penne, Michael A."Substance Use Disorder Among Older Adults in the United States in 2020." *Addiction* 104(1) (2009): 88–96.

Han, Beth, Compton, Wilson M., Jones, Christopher M., and Cai, Rong. "Nonmedical Prescription Opioid Use and Use Disorders Among Adults Aged 18 Through 64 Years in the United States, 2003–2013." *JAMA.* 314(14) (2015): 1468–78.

Harrell, Adele V. "Validation of Self-Report." In *Self-Report Methods of Estimating Drug Use: Meeting Current Challenges to Validity,* edited by Beatrice Rouse, Nicholas Kozel, and Louise Richards, 12–21. NIDA Research Monograph 57. Rockville, MD: National Institute on Drug Abuse, 1985.

Harrison, Lana D., Martin, Steven S. Enev, Tihomir, and Harrington, Deborah. *Comparing Drug Testing and Self-report of Drug Use Among Youths and Young Adults in the General Population.* Office of Applied Studies Methodology Series M-7. Rockville, MD: Substance Abuse and Mental Health Services Administration: 2007.

Hedden, S., Gfroerer, J., Barker, P., Smith, S., Pemberton, M., Saavedra, L., Forman-Hoffman, V., Ringeisen, H., and Novak, S. *Comparison of NSDUH Mental Health Data and Methods with Other Data Sources.* CBHSQ Data Review. Rockville, MD: Substance Abuse and Mental Health Services Administration, March 2012.

Hill, Craig A., Dean, Elizabeth, and Murphy, Joe. *Social Media, Sociality, and Survey Research.* New York: John Wiley and Sons, 2013.

Hoffman, John, Brittingham, Angela, and Larison, Cindy.*Drug Use Among US Workers: Prevalence and Trends by Occupation and Industry Categories.* Rockville, MD: SAMHSA, 1996.

House of Representatives. Conference Report to accompany H.R. 2264. November 7, 1997.

Institute of Medicine. Research and Service Programs in the PHS: Challenges in Organization. *Committee on Co-Administration of Service and Research Programs of the National Institutes of Health, the Alcohol, Drug Abuse, and Mental Health Administration, and Related Agencies.* Washington, DC: The National Academies Press, 1991.

Institute of Medicine. *Federal Regulation of Methadone Treatment.* Edited by Richard A. Rettig and Adam Yarmolinsky. Committee on Federal Regulation of Methadone Treatment, Division of Biobehavioral Sciences and Mental Disorder. Washington, DC: National Academy Press, 1995.

Johnson, Robert, Gerstein, Dean, Ghadialy, Rashna, Choy, Wai, and Gfroerer, Joseph. *Trends in the Incidence of Drug Use in the United States, 1919–1992.* Rockville, MD: SAMHSA, 1996.

Johnson, Robert, Hoffman, John, and Gerstein, Dean. *The Relationship Between Family Structure and Adolescent Substance Use.* Rockville, MD: SAMHSA, 1996.

Johnson, Timothy P., ed. *Handbook of Health Survey Methods.* Hoboken, NJ: John Wiley and Sons, 2014.

Johnston, Lloyd D., Miech, Richard A., O'Malley, Patrick M., Bachman, Jerald G., and Schulenberg, John E. "Use of Alcohol, Cigarettes, and Number of Illicit Drugs Declines among US Teens." Ann Arbor: University of Michigan News Service, December 16, 2014.

Kalton, Graham. "Compensating for Missing Survey Data." *Research Report Series I.* Institute for Social Research, 1983.

Kennet, Joel and Gfroerer, Joseph eds. *Evaluating and Improving Methods Used in the National Survey on Drug Use and Health.* Office of Applied Studies Methodology Series: M-5. Rockville, MD: SAMHSA, 2005.

Kessler, Ronald C., Barker, Peggy R., Colpe, Lisa J., Epstein, Joan F., Gfroerer, Joseph C., Hiripi, Eva, Howes, Mary J., Normand, Sharon-Lise T., Manderscheid, Ronald W., Walters, Ellen E., and Zaslavsky, Alan M. "Screening for Serious Mental Illness in the General Population." *Archives of General Psychiatry* 60 (2003): 184–189.

Kroutil, Larry A., Vorburger, Michael, Aldworth, Jeremy, and Colliver, James D. "Estimated Drug Use Based on Direct Questioning and Open-ended Questions: Responses in the 2006 National Survey on Drug Use and Health." *International Journal of Methods in Psychiatric Research* 19(2) (2010): 74–87.

Kurlansky, Mark. *1968: The Year that Rocked the World*. New York, NY: Random House, 2004.

Kuzmarov, Jeremy. *The Myth of the Addicted Army: Vietnam and the Modern War on Drugs*. Amherst: University of Massachusetts Press, 2009.

Lee, Martin A. *Smoke Signals: A Social History of Marijuana–Medical, Recreational, and Scientific*. New York, NY: Simon and Schuster, 2012.

Lipari, Rachel, Kroutil, Larry A., and Pemberton, Michael R. *Risk and Protective Factors and Initiation of Substance Use: Results from the 2014 National Survey on Drug Use and Health*. NSDUH Data Review, CBHSQ. Rockville MD: SAMHSA, October 2015.

Miech, Richard A., Johnston, Lloyd D., O'Malley, Patrick M., Bachman, Jerald G., and Schulenberg, John E. *Monitoring the Future National Survey Results on Drug Use, 1975–2014: Volume I, Secondary School Students*. Ann Arbor: University of Michigan, Institute for Social Research, 2015.

Miller, Judith Droitcour, and Cisin, Ira H. *Highlights from the National Survey on Drug Abuse*. Rockville, MD: National Institute on Drug Abuse, 1980.

Miller, Judith Droitcour, Cisin, Ira H., Gardner-Keaton, Hilary, Harrell, Adele V., Wertz, Philip W., Abelson, Herbert I., and Fishburne, Patricia M. *National Survey on Drug Abuse: Main Findings 1982*. Rockville, MD: National Institute on Drug Abuse, 1983.

Miller, Judith. "The Nominative Technique: A New Method of Estimating Heroin Prevalence." In *Self-Report Methods of Estimating Drug Use: Meeting Current Challenges to Validity*. Edited by Beatrice Rouse, Nicholas Kozel, and Louise Richards, 104–124. NIDA Research Monograph 57. Rockville, MD: NIDA, 1985.

Morgan, H. Wayne. *Drugs in America: A Social History, 1800–1980*. Syracuse, NY: Syracuse University Press, 1981.

Muhuri, Pradip K. and Gfroerer, Joseph. "Substance Use Among Women: Associations With Pregnancy, Parenting, and Race/Ethnicity." *Maternal and Child Health Journal* 13(3) (2009): 376–85.

Muhuri, Pradip and Gfroerer, Joseph."Mortality Associated with Illegal Drug Use among Adults in the United States." *American Journal of Drug and Alcohol Abuse* 37(3)(2011):155–64.

Muhuri, Pradip K., Gfroerer, Joseph C., and Davies, M. Christine. *Associations of Nonmedical Pain Reliever Use and Initiation of Heroin Use in the United States*. CBHSQ Data Review. Rockville, MD: Substance Abuse and Mental Health Services Administration, Center for Behavioral Health Statistics and Quality, August 2013.

Musto, David F. "Opium, Cocaine and Marijuana in American History." *Scientific American* (July 1991): 40–47.

Musto, David F. and Korsmeyer, Pamela. *The Quest for Drug Control: Politics and Federal Policy in a Period of Increasing Substance Abuse, 1963–1981*. New Haven, CT: Yale University Press, 2002.

National Academies of Sciences, Engineering, and Medicine. *Innovations in Federal Statistics: Combining Data Sources While Protecting Privacy*. Washington, DC: The National Academies Press, 2017.

National Archives and Records Administration. *Statistical Information about Fatal Casualties of the Vietnam War*, Electronic Records Reference Report. www.archives.gov/research/military/vietnam-war/casualty-statistics.html.

National Commission on Marihuana and Drug Abuse. *Marihuana: A Signal of Misunderstanding. March 1972.* Washington, DC: US Government Printing Office, 1972.

Drug Use in America: A Problem in Perspective. March 1973. Washington, DC: US Government Printing Office, 1973.

National Research Council. *Informing America's Policy on Illegal Drugs.* Committee on Data and Research for Policy on Illegal Drugs. Edited by Charles F. Manski, John V. Pepper, and Carol V. Petrie. Washington, DC: National Academies Press, 2001.

National Research Council. *Principles and Practices for a Federal Statistical Agency, Third Edition.* Committee on National Statistics. Edited by Margaret E. Martin, Miron L. Straf, and Constance F. Citro. Washington, DC: National Academies Press, 2005.

National Research Council. *Nonresponse in Social Science Surveys: A Research Agenda.* Edited by Roger Tourangeau and Thomas J. Plewes. Panel on a Research Agenda for the Future of Social Science Data Collection, Committee on National Statistics. Division of Behavioral and Social Sciences and Education. Washington, DC: The National Academies Press, 2013.

National Science Foundation Advisory Committee for the Social, Behavioral and Economic Sciences Subcommittee on Advancing SBE Survey Research. *The Future of Survey Research: Challenges and Opportunities.* A Report to the National Science Foundation Based on Two Conferences Held on October 3–4 and November 8–9, 2012. National Science Foundation, May, 2015. www.nsf.gov/sbe/AC_Materials/The_Future_of_Survey_Research.pdf.

Nixon, Richard. "Remarks on Signing the Comprehensive Drug Abuse Prevention and Control Act of 1970," October 27, 1970. Online by Gerhard Peters and John T. Woolley, *The American Presidency Project.* www.presidency.ucsb.edu/ws/?pid=2767.

Nixon, Richard. "The President's News Conference," March 24, 1972. Online by Gerhard Peters and John T. Woolley, *The American Presidency Project.* www.presidency.ucsb.edu/ws/?pid=3356.

Nixon, Richard. "Remarks at the First National Treatment Alternatives to Street Crime Conference," September 11, 1973. Online by Gerhard Peters and John T. Woolley, *The American Presidency Project.* www.presidency.ucsb.edu/ws/?pid=3958.

Office of Management and Budget. *Statistical Policy Directive Number 2: Standards and Guidelines for Statistical Surveys.* Washington, DC: OMB, September 2006.

Statistical Policy Directive No. 4: Release and Dissemination of Statistical Products Produced by Federal Statistical Agencies. Washington, DC: OMB, 2008.

Guidance for Providing and Using Administrative Data for Statistical Purposes. Memorandum M-14-06. Washington, DC: OMB, 2014.

Office of Management and Budget. *Statistical Programs of the United States Government, Fiscal Year 2017.* Washington, DC: OMB, 2017.

Office of National Drug Control Policy. *National Drug Control Strategy, September 1989*. Washington, DC: US Government Printing Office, 1989.

National Drug Control Strategy, 1993. Washington, DC: US Government Printing Office, 1993.

Pew Research Center. "News Use Across Social Media Platforms 2016," May, 2016.

Pierson, Steve. "ASA, with Statistical Community, Watching Carefully for the Integrity of Federal Statistical Data." *ASA Science Policy blog, Feb* 2, 2017. http://community.amstat.org/browse/blogs.

Pollin, Teresa and Durell, Jack. "Bill Pollin Era at NIDA (1979–1985)." *Drug and Alcohol Dependence* 107 (2010): 88–91.

Prewitt, Kenneth. *What is Your Race? The Census and our Flawed Efforts to Classify Americans*. Princeton, NJ: Princeton University Press. 2013.

Reagan, Ronald. "The President's News Conference," March 6, 1981. Online by Gerhard Peters and John T. Woolley, *The American Presidency Project*. www.presidency.ucsb.edu/ws/?pid=43505.

"Radio Address to the Nation on Federal Drug Policy," October 2, 1982. Online by Gerhard Peters and John T. Woolley, *The American Presidency Project*. www.presidency.ucsb.edu/ws/?pid=43085.

Rettig, Richard A. and Yarmolinsky, Adam eds. Federal Regulation of Methadone Treatment. *Committee on Federal Regulation of Methadone Treatment. Institute of Medicine*. Washington, DC: National Academy Press, 1995.

Robins, Lee N. "Vietnam Veterans' Rapid Recovery from Heroin Addiction: A Fluke or Normal Expectation?" *Addiction* 88 (1993): 1041–1054.

Ryan, Heather, Trosclair, Angela and Gfroerer, Joseph. "Adult Current Smoking: Differences in Definitions and Prevalence Estimates–NHIS and NSDUH, 2008." *Journal of Environmental and Public Health*, 2012.

Rydell, C. Peter and Everingham, Susan S. *Controlling Cocaine: Supply Versus Demand Programs*. Santa Monica, CA: RAND, 1994.

Schober, Susan and Schade, Charles eds. *The Epidemiology of Cocaine Use and Abuse. NIDA Research Monograph Number 110*. Rockville, MD: ADAMHA, 1991.

Sirken, Monroe G. "Network Surveys of Rare and Sensitive Conditions," In *Advances in Health Survey Research Methods. National Center for Health Statistics Research Proceedings Series* (1975): 31–32.

Streatfeild, Dominic. *Cocaine: An Unauthorized Biography*. New York, NY: Picador, 2001.

Substance Abuse and Mental Health Services Administration, Office of Applied Studies. *Substance Abuse in States and Metropolitan Areas: Model Based Estimates from the 1991–1993 National Household Surveys on Drug Abuse, Summary Report*. Rockville, MD: SAMHSA, 1996.

The Development and Implementation of a New Data Collection Instrument for the 1994 National Household Survey on Drug Abuse. Rockville, MD: SAMHSA, 1996.

National Household Survey on Drug Abuse: Main Findings 1995. NHSDA Series H-1. Rockville, MD: SAMHSA, 1997.

Preliminary Results from the 1996 National Household Survey on Drug Abuse. NHSDA Series H-3. Rockville, MD: SAMHSA, 1997.

Analyses of Substance Abuse and Treatment Need Issues. Analytic Series A-7. Rockville, MD: SAMHSA, 1998.

Summary of Findings from the 1999 National Household Survey on Drug Abuse. NHSDA Series H-12. Rockville, MD: SAMHSA, 2000.

Summary of Findings from the 2000 National Household Survey on Drug Abuse. NHSDA Series H-13. Rockville, MD: SAMHSA, 2001.

Development of Computer-Assisted Interviewing Procedures for the National Household Survey on Drug Abuse. Methodology Series M-3. Rockville, MD: SAMHSA, 2001.

National and State Estimates of the Drug Abuse Treatment Gap: 2000 National Household Survey on Drug Abuse. NHSDA Series H-14. Rockville, MD: SAMHSA, 2002.

Summary of Results From the 2002 National Survey on Drug Use and Health. NHSDA Series H-22. Rockville, MD: SAMHSA, 2003.

The NSDUH Report: Seasonality of Youth's First-Time Use of Marijuana, Cigarettes, or Alcohol. Rockville, MD, June 4, 2004.

Results from the 2004 National Survey on Drug Use and Health: National Findings. NSDUH Series H-28. Rockville, MD, 2005.

The NSDUH Report: Impact of Hurricanes Katrina and Rita on Substance Use and Mental Health. Rockville, MD: SAMHSA, January 31, 2008.

Substance Abuse and Mental Health Services Administration, Center for Behavioral Health Statistics and Quality. *Comparing and Evaluating Youth Substance Use Estimates from the National Survey on Drug Use and Health and Other Surveys.* Methodology Series M-9. Rockville, MD: SAMHSA, 2012.

The NSDUH Report: Accessing National and State Data from the National Survey on Drug Use and Health. Rockville, MD: SAMHSA, August 3, 2012.

The NSDUH Report: Trends in Insurance Coverage and Treatment Utilization by Young Adults. Rockville, MD: SAMHSA, January 2013.

Results from the 2013 National Survey on Drug Use and Health: Summary of National Findings. NSDUH Series H-48. Rockville, MD: SAMHSA, 2014.

Teich, Judith L. and Pemberton, Michael R. "Epidemiologic Studies of Behavioral Health Following the Deepwater Horizon Oil Spill: Limited Impact or Limited Ability to Measure?" *Journal of Behavioral Health Services and Research.* 42(1) (2015): 77–85.

Tourangeau, Roger, Rips, Lance, and Rasinski, Kenneth. *The Psychology of Survey Response.* Cambridge, UK: Cambridge University Press, 2000.

Turner, Charles, Lessler, Judith, and Gfroerer, Joseph eds. *Survey Measurement of Drug Use: Methodological Studies.* Rockville, MD: NIDA, 1992.

US Department of Health and Human Services. "Closing the Treatment Gap: A Report to the President of the United States." September 2001.

Voss, Harwin L. and Clayton, Richard R. "Stages in Involvement with Drugs." *Drug and Alcohol Abuse in Children and Adolescence. Pediatrician* 14 (1987): 25–31.

Wang, Kevin, Kott, Phil, and Moore, Andrew. *Assessing the Relationship Between Interviewer Effects and NSDUH Data Quality.* Prepared for the Substance

Abuse and Mental Health Services Administration under Contract Nos. HHSS283200800004C and HHSS283201000003C. Research Triangle Park, NC: RTI International, October 24, 2013. www.samhsa.gov/data/sites/default/files/NSDUH-IntEffects2013/NSDUH-IntEffects2013.pdf.

The White House, Office of the Press Secretary, Fact Sheet on the President's National Drug Control Strategy, February 12, 2002.

Whitford, Andrew B. and Yates, Jeff. *Presidential Rhetoric and the Public Agenda: Constructing the War on Drugs*. Baltimore, MD: The Johns Hopkins University Press, 2009.

Wish, Eric D. "US Drug Policy in the 1990s: Insights from New Data from Arrestees." *International Journal of the Addictions* 25 (3)(1990): 377–409.

Woodward, J. Arthur, Bonett, Douglas G., and Brecht, M. Lynn. "Estimating the Size of a Heroin-Abusing Population Using Multiple Recapture Census." In *Self-Report Methods of Estimating Drug Use: Meeting Current Challenges to Validity*. Edited by Beatrice Rouse, Nicholas Kozel, and Louise Richards, 158–171. NIDA Research Monograph 57. Rockville, MD: NIDA,1985.

Wright, Douglas, Gfroerer, Joseph, and Epstein, Joan. "Ratio Estimation of Hardcore Drug Use." *Journal of Official Statistics* 13(4) (1997): 401–416.

Index